Journey Around an Extraordinary Planet

by Johnny Dolphin

First edition.
Published by Synergetic Press, Inc.
Post Office Box 689, Oracle, Arizona 85623

Edited by Kathelin Hoffman
Cover illustration by Ralph Steadman
Book design by Kathleen Dyhr

ISBN 0 907791 23 9

Printed in the United States of America by Arizona Lithographers, Tucson.

Table of Contents

An adventurer without expeditions,
An artist without performances,
A scientist without experiments,
A philosopher without a critique,
A manager without a project,
A mystic wanders a journey without a road.

At night the gravel is soft as roses,
The dry bread and water sweet as chocolate,
Solitude intenser than high politics,
Starlight more penetrating than poetry,
Death becomes the only game in town,
And timing calls all the shots.

The Magic Room 1963

I HAD ESCAPED America, but I had not found the truth. I sat in the Zocco Chico, sipping the sweet minted tea, lipful after lipful, watching the watchers and the watched, being watched.

I lived on tangerines, dates, figs, round loaves of thin bread, goat cheese, a daily bowl of soup, and mint teas. I paid twenty dollars a month for a five story house deep in the Medina, one room on each story, with a sturdy but wonderfully irregular staircase culminating in a balcony from which on a clear day I looked for and saw impressions my mind formed into the imperial word "Gibraltar". For the first time in my life I lived in architecture rather than a building.

Beni Sidi waved me over to his doorway on the block past the park from which you could see the Mediterranean just past the Café Paris gracefully perched at the end of the long curve up from the Medina. Two times a day I walked up to the Café Paris for a mint tea and European or American expatriate conversation. I never walked on to the famous expatriate bar. The alcohol scene did not interest me. I didn't wish in any way to suppress cerebral activity. Hashish for that period became my only experimental tool to supplement watching. Watching my emotions and behavior. Watching others' behavior. Watching made a good first step.

"What do you want?"

"I don't know."

"How can I get it for you if you don't know what you want?"

My life halted, my thoughts stopped. Engraved in temporarily crystallized sensitive energy I saw the words "I don't know what I want". I had not searched for truth up to that moment, but truth hit me. I did not know what I wanted, or if I wanted anything. I did not exist.

Joe Madison existed. He had made a happy simple life for himself in Tangier. But here a man stood who asserted he could get me what I wanted, if only I knew what I wanted, and if only I had me instead of Joe Madison having me.

I believed then that Beni Sidi could get it for me if I knew what it was and I believed him all the time I lived in Tangier. I finally bought a little round wooden table from one of his friends and twelve cushions from another and a mattress from yet another. Also he got me a meeting with the most beautiful Berber whore in Tangier. "She makes love like a snake." Her green eyes glittered and glowed, her neck arched, her perfect pinked fingernails lightly indented my wrist. But I found five minutes of talk with her satiated indeed overflowed my hyperbolic sexuality and I walked out after the price had been fixed. I found it interesting to a very fine degree that I did not make love to this woman. Already by that time I could "breathe in" a human being and I did breathe her in, and, before "breathing her out", imprinted a trace immortal as my life.

Beni Sidi showed me the workings of the life of Tangier, the little round tables being nailed together, leather being suppled, the couscous turned into cuisine, and noted to me the availability and cost of various girls, boys, musics, drugs, and scenes.

Something moved inside me, melodies of outwardly soundless music yet as clear as if I owned a metaphysical dog who could hear inner octaves beyond the merely human ear. The sounds of the Berbers rising in faint synchronic waves from the crowded street traffic of life began to order my hearing in pre-Bachian modes from which the rarest melodies could be differentiated in a majoom ambienced midnight makam.

Day after day I sat, elliptic, erect and for hours, at my chair in the Zocco Chico. Finally, I could see the whole extraordinary daily procession of images as being generated from an immense field of force that itself possessed no color. I could separate the film from the projector, the movie house, and the audience. But I did not make any effort to change the film because I didn't want any other than the Zocco Chico film. I let every day run the film and every night rewind the film. I studied every frame and found nothing that made me wish to stop the film at any point. I studied the montage within the frames to see what the discontinuities said about the director's vision.

I began to see how each day spliced together a slightly different film as if Eisenstein had decided never to present a finally finished product and that this mode was the true finished product, ever the same and ever new. One day a dervish strode through the Zocco in a cloud of Allah that kept the closest people three feet away from him. I knew he bore a message from my future, calling me to follow him, but the film held me. I wanted all of the marrow from those image bones.

One day a curved-nose, green-eyed, thatch-haired tall thin Englishman about thirty, coiled beside me and ordered a mint tea. A man two days later handed me a photograph that he had carefully developed of Terry and I sitting together. He said he recognized the moment's importance and knew that I could use a record.

We sat for a long time side by side, our mint teas on the same small table, gazing at the scene. He wore a jelaba. Time passed and we understood we both "saw". My first week in Tangier had ended and I had till then met only Beni Sidi by name. "Going up to the Kief Cafe. Want to come?" "Sure." Though I didn't know where or what he meant. He rose, lanky-boned, and smoothly moved around the corner into a side street that led up the hill toward the Casbah.

"This house belongs to Barbara Hutton, the millionairess." Abruptly above and to the left of the Hutton house sat the Kief Cafe facing out over the space above Tangier. Mats of subtle red and yellow, a broad flat table in the rear room. One room open to the terrace had four small tables side by side. On the terrace three tables. Three telephone wires cut diagonally across the city that I already loved. Mid-November, the rains had not started, a few white clouds emphasized blueness of sky.

After ordering the mint teas, Terry brought out his kief bag. He began deftly and with luminous attention to separate seed from the dried leaves. The seeds grew in size for me until they became as large as peas. He worked like a goldsmith. Then he laid out two pipes, wooden, carved, painted.

I had gone on two peyote trips and for seven days had done a small amount of hashish, but never before had I seen anyone who knew what to do, exactly. I abandoned myself to a master. We began to smoke the pipe at about one o'clock.

The white cloud underwent minute changes in the sky with three lines unchanging across it. The minute changes resulted in complete symbolic transformations.

I reached to a place past even the force that generated the montages that generated the film at Zocco Chico or in the sky above me. I went nowhere. Not like at night I disappearing into sleep. I went nowhere and stopped and did not disappear.

The cloud above me loomed majestic, dazzling, the sky gleamed blue. I felt life begin anew. I had gone to a place where everything had been taken from me that held me to a position, a time, a name, a fate. The incalculable had occurred. "What time is it?" I asked Terry.

"Five-forty-five."

So time did not rule the entire world. I had been to nowhere, I had experienced nothing, everything had changed, and this had occurred in no time, and upon return from the incalculable, time had ticked away over four hours in its world. Incommensurable.

"Let's go get some soup from that old Moroccan soldier by the gate."

The Moroccan filled two wooden bowls with hot soup and pieces of lamb. "I fought in Spain. We finally fought our way into Barcelona. We looted and plundered. We are great fighters." He laughed pleasantly. "With my retirement and my plunder and this little shop I do very good."

Terry and I looked at each other then for the first time. His great green eyes lived like two subaqueous mammals hooked to an electric eel that turned them on and off.

"So much for the agonies of Western Man, left, right, left, right," I said.

"Too much," said Terry.

The missing four hours interested me more than the entire history of Europe. Plato, Eckhart, Galileo, Marx, Nietzsche: none of them had anything to say about nothing.

I wanted to return to that nowhere again and I wanted to be completely conscious when nowhere.

Terry and I started the Magic Room two nights later. He hung the sides of a large room, fifteen feet by twenty feet by ten feet high, with Berber rugs. The stripes on the rugs all ran irregularly parallel. He hung the rugs so that the stripes ran vertically. Then he laid the floor with Berber rugs. I saw that every line, represented a life and that ordinarily these lives never met another life from beginning to end even as they made a rug together, and that every rug represented a group of lives that never even touched another group of lives except from time to time as stronger forces rearranged their location to change the overall vibrations.

Hamid, a handsome Arab, Walid the mute, Terry, and myself formed the core along with Lita Chayevsky. Lita had told me she would probably meet me there when she left New York a month before I did. She ran up to me, laughing, at the Café Paris. "See, I got here!" We had both changed too much for that. But some of the connection from the peyote night remained. I remembered her hair turning to dry feathers in my hands beneath a killer moon. I gave up my five story building and moved into the room on the roof of Terry's building, introduced Lita, and she became Terry's girlfriend.

Each night in the Magic Room we'd also invite one or two new people that seemed to have the talent to do it and the being to survive.

Each one would concentrate, projecting his inner scene. The one with the most power would make the scene that would take over the night in the Magic Room. That one would have made the greatest magic. I learned how to measure power. Terry, lean, deft and poised, prepared the kief from the dried plants, carefully selected from the Berber women's stocks. Then he would pass out majoom cookies.

We sat backs to the wall in silence focussing on making the scene appear. In one I heard Walid, the mute, screaming, "Let me out! Let me out!" His eyes burned like a man newly sentenced to life imprisonment. Terry and Hamid became one glance which became tensile, material, then alive, as their two I's danced out upon that high wire that their live bodies lavished their energies upon creating and maintaining. Lita and Mark, two bright six year old Jewish kids from Shtetl immigrants, played pat-a-cake on the sidewalks of the Lower East Side. My head rolled off my right shoulder and sat on the floor, taking it all in while I watched myself become a contemplative head with no body to care for or react to. This scene had not been the one I projected nor did anyone else claim it. The great magic scenes came from an undiscoverable magician.

Gradually I realized the Magic Room with blue ceiling and

Berber rugs enclosed all of Tangier from the Casbah to the Medina to the cafés in front of the ferry, the front beach, and the two streets up to the Café Paris. But when I walked on across the street past the Café Paris, always filled with its busy cars, somehow I had walked past the Magic Room. After awhile I never walked across that street. I needed to be reborn but I could not die because I had not waked up to my total life, only to moments, hours, days, and I did not want to sleep through the same life twice. Day after day Joe Madison, six feet one, tough, etcetera, became more helpless. His weight dropped. Night after night he said goodnight to Terry and Lita looking fondly at each other, or Lita washing Terry's long blond hair, and then would climb to the roof where he slept each night a little more highly curled up.

Joe Madison of Oklahoma disappeared and I woke up in a new house in Tangier with the loveliest parents that I could imagine. They did not come from old straight shooting hardworking puritan frontier stock. Subtle, spiderwebbing, they didn't work; pleasure-loving, esthetic, they explored sensation and emotion not space and intellect.

Lita's short curly hair framed a face of halcyon ecstasy. "I only slept with eleven men in the two months getting here. I don't regard that as excessive for a modern girl traveling. And besides at the fountain in Naples, one of them kept sticking flowers between my toes until I just had to stop reading my book."

She painted broken, jagged glossy elongated planes colliding with distorted circles in a purplish-golden atmosphere. She wore a dress, gold earrings, sandals, and panties. She moved like the changing winds of April. Her brown eyes looked out at the world like those of a summer swimmer floating on her back on a hot day.

Terry labored for the perfect pipe of kief and the perfect cookie of majoom. Terry labored to become the perfect observer. Terry got the unreliable cheap Chinese batteries that did enable us to hear what he said we must hear by candlelight, the Beatles, and Ginsberg doing *Howl.* Terry wanted to turn all London on and later helped start the process with street acid together with his tall, thin-nosed call-girl friend from Chelsea. Terry could talk about the fine points of sentence structure and the power of paragraphs. He had written a novel of which he had no copy available and had been in prison in England once on a drug count.

I decided to grow up with these parents. Lita and Terry. I loved them, admired them. One night I felt like growing up. I wanted Mamma to come to the roof with me. I made hierophantic gestures to her at the bottom of the stairs. She shook her head, and looked at Terry whose green eyes gazing back, waved electrically in a field of seaweed. She stayed with Terry but I understood and slept happily.

The next morning we three gathered as usual on the roof for tea and tangerines. The white roofs of Tangier gleamed down the hillside, city beyond any compare of mine, with waiting citizens who inhabited possibility. I excitedly looked forward to telling Terry and Lita my secret. That I had become a new man with new parents, and who my new parents were.

"You will have to leave this morning," Terry said.

I beamed at both of them, happy as a child. "Okay, I'll be back this afternoon."

"I mean you have to move out of the house, now."

I looked at Terry, then Lita, then Terry with amazement. I loved them. I had introduced them.

"Can't you hear me? You have to leave, now."

"Why, Terry, why?"

"Because I can't get it out of my mind that you used to sleep with Lita, that you're on the roof above us, I can't go on with it. I want you completely out of my life, out of her life, and out of my house."

I couldn't say anything. Kicked out of home. New parents, new world. I had jubilantly left miserable home in the first life. Now kicked out of home, miserably, in my second, jubilant, life.

Down to the public shower house run by the old henna-haired lady. Showered, still dismayed, with aching body, walked to the lobby when Terry entered. Murderous strength filled me. Walking up to him I suddenly saw how scrawny and weak he was. I could kill him quickly with a blow at his bulging Adam's apple. He deserved it. He had kicked me out. I had lost my new life almost before it had begun. I raised my fist to kill him.

I reached that nowhere again. I struggled to remain conscious. I saw far away and hazily what would happen if I killed him, what would happen if I didn't. How decisively life lies in wait at crossroads. My body commenced to change under the circulation of all the energies that in the past would have gone into hitting, kicking, yelling, repressing, suppressing, expressing, impressing, pressing.

My body rigid as a rock, my hand gently, fully extended on his thin neck when I returned. His phosphorescent green eyes impaled on their stack of sexual energy regarded me with the perfect objectivity of the accomplished writer, constating his inner sensations together with those coming from outside.

"I intended to kill you," I said, "but now I've decided not to waste my life."

"I've always wanted to meet a man who wanted to kill me," he said. "and I always wondered if I would survive."

I strode past him rich in the knowledge no parents would ever turn out significantly different. One huge heartache I never knew existed stopped. The entrance to the home I wanted started in the nowhere, not with the parents and their unconscious random shuffling of myriad genes and memes.

At the Zocco Grande I began to bargain with a man hustling a beautiful jelaba at 90 dirhams. At last I could get into it. The fight, the rhythm, the passion.

"Fifteen."

"OK, OK for you eighty-five. This beautiful jelaba. Made for you. Try it on."

It fit perfectly. Subtle vertical stripes, alternating grey and whitish grey. Nothing was thought out. My pulses simply matched his. My movements matched his. My interest in dirhams matched his. From this moment I was going to live as an Arab. I connected to the undiscovered magician in the Magic Room. The sky broke open on Zocco Grande and the face of Allah glanced back at every glance. The battle took place in no place.

"Thirty-five." Maybe fifteen minutes had passed to reach this point.

"Forty! Forty! Forty! It has cost me forty I swear by Allah I am not making a dirham, nothing, you have stolen my day. Forty!" His eyes glared at me, murder, frenzy, determination. The Moroccans massed around us, all of their eyes giving out energy, taking in energy.

"Thirty-five! Thirty-five!" Now take it or leave it. I wanted to buy that jelaba, I wanted to buy that jelaba and not pay one extra dirham for it or it would be forever soiled by weakness. We circled each other.

"Thirty-eight!" "Thirty-five!" Finally at "Thirty-six," he screamed, threw down the jelaba in front of me and stuck out his open palm, fingers taut. I picked up the jelaba, dropped it on over my head and counted him out thirty-six dirhams. I pulled the magnificent cowl of the jelaba up over my head. I already wore the yellow open-heeled slippers of Morocco. The crowd had been cheering and clapping since he had flung down the jelaba at my feet. I breathed them in, imprinted, and breathed them out. My right arm waved wildly at them as my feet strode out of Zocco Grande.

At last I could leave the Magic Room and move out across North Africa under the skies of Allah, the merciful, the compassionate, the powerful, the subtle, the beautiful, the learned, the subsistent, the truth. That night the winter rains commenced in earnest. The world turned thousands of shades of grey. I drifted down the streets at last one among many, toward the bus stand. First, Fez, and then through whatever lay between Tangier and Upper Egypt.

A beggar called to me from in front of a mosque. Looking at the beggar told me that I could no longer lose the battle of life. At worst I could sit and beg in front of a mosque. I would be more likely to find what I wanted there than anywhere in Europe-America. But deeper knowledge, I felt, lay in the ruins of Imhotep and Amenhotep than in the atmosphere of the spices which evoked the magic carpet. I wanted to learn how to weave that carpet for myself, and how to move upon it at will between nowhere and anywhere in those gaps I had found out how to see between the montaged photographs that made up the illusion of the movie Zocco Chico.

The Tombs of Knowledge

In a city in Algeria built on both sides of a steep gorge, I checked my pockets shortly after leaving the hospitality of a house whose French owners had invited me to sleep overnight. I had left my watch and reading glasses.

Curses at my forgetfulness, panic lest they could not be reobtained, but I did not move for several minutes. They were my last western-made possessions except for my khaki pants, scissors, journal, wallet, passport, and one thousand dollars of traveler's checks out of which I figured I'd spend forty dollars a month, about the average income of an adult inhabitant of the region. I had only a shirt change, tangerines, dried figs, and dates in my Berber-made straw shoulder bag. I decided to do without the watch and the glasses. I had worn a watch and glasses while reading for twenty years, since I was fourteen.

I decided not to need them anymore. I wished to make higher energies and form-seeing out of my organism, not out of the apprehension of a few more phenomenological details. Days of exploration followed as I began to assimilate the time and energy devoted to the watch and the glasses into existing on the moment-by-moment watch for "that state" to come again.

A mad lover, I staggered across Africa desperate for another embrace, even a touch, even a glance.

In Cairo, the city of Mars, of the Pharaohs, of the Fatimids, of Napoleon, the Empire, and always of aspiration, I galloped the sands beyond the river, stirrup to stirrup with an Arab who rented Arab horses. The staggered pyramid of Sakkarah and the sacred bulls lying in state in ranks of tombs beneath the ground spoke to me more eloquently than the Sphinx and the Great Pyramid whose knowledge I felt could be assessed and did assess as potent knowledge of individual destiny and the social machine. But the knowledge that could create an empire, a religion, and a culture of pyramids and animal forms, this knowledge I saw, came from minds that had mastered the invisible forces that create the images of powers and of possibilities. Hidden, hidden, deeply hidden the wisdom of Imhotep.

The long meditative train ride to Thebes and the New Empire. The Old Empire destroyed, the age of Taurus the Bull discarded, someone had built the great Way of the Ram of Aries at Thebes, carved the reminiscent nostalgia-laden papyri columns of Karnak, and buried the kings in rock-cut tombs, just as before the bulls had been buried. Someone had known how to start all over again with new forms for a new age after the savage destructions of the Hyksos and Time.

Paper and Wool, instead of Rock and Leather. The king below and the animal above, instead of the king above and the animal below. Many columns, each equal, instead of the one pointed pyramid. Paintings of the complete life, rather than the austere ship of the dead. Such transvaluation of values on such a scale, in the most powerful empire of those centuries, that those who accomplished this must indeed have known how to enter the realms which by now not only I wanted but desired and wished, to be united with.

The Sheik's name had been given me by a Swedish artist, Helga, whom I had met at the Cairo Museum.

"The Sheik's place is near the Colossi of Amenhotep. Tell him Helga sent you." She recorded the paintings on the cool earthen houses in the Nubian villages soon to be flooded by Aswan Dam. The Nubians would be relocated into hot concrete boxes on the desert above the lake.

"I make part of my money doing this, and the other half of the year I weave designs into the wool from my sheep near Bad Gotesburg. You could visit me there if you like."

A slim, graceful woman of twenty-eight, with sea-green eyes, with fingers that could do anything short of the rigorous nuances of Eastern dancers. Till four months ago she would have been my heart's desire.

Not that I didn't feel toward her exactly what I would have felt for her. But feelings no longer controlled my organism. I was borne onward beyond woman; not anti-woman, in fact, I could appreciate the rarity, the goodness, even the splendor of Helga more than possible before, but something different controlled my direction.

"I see," she said as we sipped Turkish coffee on a small square wooden table in a crowded hookah-filled café, next to a pair of absorbed backgammon antagonists. "Definitely you should see the Sheik. Maybe I will also be there when you arrive."

The Sheik ran a colony for Egyptian artists, but some few Westerners, including Helga, and upon presentation of a slip signed by her, myself, were allowed to stay. The Sheik, lean and old, emanated power and decision.

"Why do you wish to stay here?"

"I have been told that staying here will assist me to discover what I am looking for."

After a week I looked at Helga. A beautiful woman. Across the river for a coffee at the big tourist hotel. She tells me how she slept with the old Swedish archaeologist there. "One night. It is perhaps his last chance. He did such wonderful work." I watched memories of old seductions rise in my mind but they no longer controlled me. Neither did I make any effort to fight them off. We did not take a hotel room, but we stopped to talk in the edge of the little road leading to the Sheik's compound. Sitting on the road's edge, the clear stars above, I felt the crystalline splendor of sex creep upward.

The Sheik's man arrived. "This is where you are. Leave immediately. The Sheik says you must come back immediately. The desert fills with vipers in the night. Very dangerous."

The next day on the way to the Caves of the Kings, we took a cut across the ridge and went into a small cave that we found. We kissed. Nothing happened to me. The most beautiful intelligent woman I had ever met. This, then, was not what I wanted. I had learned, somewhat, to sail with my body, not to impose fantasy in the name of romance. For the first time in my life, the realization arrived that I did not have to desperately, terrifiedly act out the passionate male animal when no passion existed. I did not have to fuck on penalty of losing my manhood.

Filled with the energy of new hope I departed for the South, for the Sudan, for the source of the Nile. Suddenly even Egypt's history suffocated me in its newness, its rawness. Where did the Sphinx come from? Where did man himself come from? How could I go any further without knowing that? South, south, south murmured the call.

That turn south looking for man, instead of on to the Middle East looking for mind, changed everything irrevocably. I had escaped not only America with its cold promises of wealth, but Egypt with all that implied of the miracles of secret knowledge about the alchemy of thought. Of the thousands of paleskins in Cairo and Thebes, I was to be the only one on any of the vehicles or ships I took or roads I walked toward the south where a dirty racial war did its daily killing. No one from any place I had ever been knew where I was; no one I met along the way had any but the vaguest idea from where I had come. I could have disappeared without a trace. That thought never occurred to me. I knew I could never lose following this certainty. I was becoming as old as history, but each morning was as fresh and delicious as an immortal youth confident of finding his daily lover.

The Chief Who Escaped Hanging

FROM WADI HAIFA which waited in Nubian tempo for the flood to arrive upstream from the Aswan Dam that would drown its way of life, I took a third-class seat in the train on the old Kitchener railroad built to destroy El Mahdi, the dervish nationalist leader who had killed Chinese Gordon, the conqueror of China. My jelaba had been traded in for an extra khaki shirt and new pair of khaki pants. The Moroccan slippers continued, my scissors used only to keep the beard and adhere to local lengths. More surprising to me than my body's giving up girls was its giving up of reading, of carrying even one book except for my small journal for two or three lines on some days. The books too had left without a struggle; no self-denial, simply a vanishing of any need to do other than follow my desire to perceive directly. Every person encountered became a teacher because all I wanted to do was to learn, to extract the secret of how they survived this tumultuous passionate world where desire always and need very often exceeded the resources available even to the quickest and most alert. I reduced my technique to travel and the way of relative poverty so though people and cultures changed, the necessity of my dealing directly with everyone who had power over a poor foreigner, to help or to hinder did not change. Without sufficient cash and with no position, one comes face to face with power, that is, the other must directly decide how much of what they will give me. And their calculation of the limits of their power together with their evaluation of me, and my organic reactions to how they carried out these operations of anger and compassion taught me bit by bit about the world, its people, and myself.

My companion on the hard wooden seat on the crowded train offered me a fat cigar, similar to his own, stuffed with hashish and marijuana rather than tobacco. We smoked companionably together and the creaking lurching train melded with the smoke inside, the night outside, the smells, sighs, and unfocused dark eyes. The unbearableness that death continually destroyed this splendor of the living smote me with intoxicating hurt.

The train braked to a metallic halt. Outside, no town, not even a house visible over the desert. Crones arrived with twigs to make a small fire over which they placed black kettles. Tea.

"We speak the most perfect Arabic of all the Arabian countries," my friend said at last in perfect English. "We were left to ourselves in Omdurman, and so our use of the language was never corrupted. I invite you to stay with me at my house upon our arrival. I will show you everything to the best of my ability. My family is one of the oldest. Look at this woman. She is a witch. We will buy our tea from her."

He said something to her in Arabic. She craned into my face. Her eyes, yellow around the pupils, darted her glance into me, seizing her emotional fish. Returning to her kettle, she popped in some herbs.

We drank slowly sitting crosslegged on the sand under the dark star-pierced sky of the mild-wintered Sudan.

In Khartoum, we quickly made our way out of the pseudo-European city and across the river to Omdurman. Through a gate in a mud wall, we stepped into a garden surrounding a rambling adobe house. After introducing his cheerful father, he showed me the hole in the ground in the pleasant outhouse, open to the breeze, and then to the cot that would be mine on the veranda so as not to disturb in any way the tenor of the harem.

After paying our respects to El Mahdi in his giant mausoleum, we went to visit my friend's second wife, Galla witch, who lived in a thatched hut in another section of Omdurman. "We all have our Muslim wives in order to make our sons, but in addition we have another, a Galla, for our pleasure and to whom we can talk to about everything to our profit." Seeing the Galla working with her jars of foul-smelling mustily fragrant ointments and her rafters of dried herbs, laughing, moving about the storm lantern lit vastness of her territory, staring at me with eyes that had never known inhibition, sizing me up, letting me size her up completely, made up my mind for me. There are two sources to the Nile: one in Ethiopia, the other in Uganda in the center of Africa. I could see

Ethiopia now in the clever grace of the witch-wife. The Blue Nile would take me perhaps to the historic source of Egypt, but I could see that I would not make it back to the real tribe, to before Egypt's conception, before the Sun God, before civilization. The Galla could not fool me; she knew too much, thousands of years of physically submitting while psychologically manipulating too-civilized men like my friend, the deputy cabinet minister; she was a symbiotic. If I had met her first upon leaving New York I would have been no match for her. Now I watched her give my friend some of "the shameful drink", his relaxation, her making a charm for his success. We traded glances over his head. I smiled so she would know I applauded her skill. She would make a relentless enemy intent upon stifling any threat. It was up the White Nile I would go, through the Sudd, the papyrus-choked waterway of crocodiles and hippos to the lands of the Nuer, Shilluk, and Dinka. From what did I spring before civilization ever conceived itself in the mind of some mighty school of shamans and warriors? All had come by way of the tribe.

The railroad ended at Kosti. Long lines of white clad men doing their prayer exercises. Not a single priest in front of them, only Mecca. With the end of the railroad had come the end of even the rumors of the death of God. Submission. Islam. The men submitted to community, to the exercise, to Allah, to the hot sun on their back as they bent forward, the women submitted to the submitted men, the land itself they left submitted to the elements, the elements submitted to the jinns, and the jinns submitted to the limitations of powers set forth when the pen first wrote upon the tablet and from that infinitesimal point exploded the universal story.

My entire organism convulsed with the wish to make the confession of faith, to join the line of men, to cease the search and give myself up to the massed emotion, the glisten, the heat, the discipline, the bare ground, the hope swinging through the air tangible as a giant sword, a hope so substantial that all modernity seemed but a despairing collection of ill-made gewgaws to mock the individual's apprehension of death, his own, his culture's, his world's.

Realizing, however, from Morocco, Tunisia, and Egypt that this purity existed only because the Mullahs here, unconcerned about whatever means it might take, had so far been successful in keeping their people from depth contact with any other forms of thought and life. Yes, submission to the Mullah was also included in this package, submission to his scholarly hatred of all other forms of manifestation, all of which, of course, also must be permitted by Allah since they, too, flourished whenever the Mullah could not coerce.

Breathless from sharing in the exaltation, but determined to leave that crystalline stasis, when the lines broke up and the men walked back to their various occupations, I strode excitedly past the passenger steamer toward the barge that the steamer would push ahead of it. Throngs of tall lean spear-carrying men had gathered in a hubbub.

Up the gangplank to the barge deck. Rush for a bunk. My toes snuggled from time to time during the next ten nights to Juba against the toes of the nearly seven foot Shilluk warrior sleeping on the bunk headed against mine. The armed Muslim Arabs immediately forced the segregation of the women from the men to the disgust of the southerns.

Then Nuba guards with rifles took up strategic positions. The spears were taken from all the tribesmen and stashed in the first and second class ship.

"These Nubas will watch us all the way to Juba," a big broad shouldered Dinka said to me. We had been standing together at the rail since the boat train had swung away from the dock. "They are Muslim now. They hold us in contempt. But watch them. They do their prayers to Mecca while the ship is turning a bend. They don't even know what direction they are going in anymore."

The Nile slid by. At night and in the morning and at noon the women brought food. An almost raw doughy bread, chicken, rice. I had obtained malaria pills in Cairo and chlorine pills together with a canteen and traded my Moroccan slippers for a pair of old tennis shoes.

The Chief, for that was his position amongst the Dinka, and I often stood together for hours on the rail watching the river, the sky, the banks, the tribesmen, the Nuba guards with the ready rifles, the Arabs hawk-like up on the deck of the ship, the captain, the helmsman, the steward who from time to time would look down upon us from the bow. Occasionally laughter would ring out, occasionally a small hubbub, but mostly silence. Silence that was in no way isolating, negative, introspective. A warm silence that grew more transparent each day. Someone would point, everyone would look. An alert silence. No one was allowed to disappear into himself.

"Hip," I said to myself. "Hip. No wonder it came to us via jazz from Africa. They don't allow you to miss a beat."

No wonder tribes never built a Sphinx; they had living ways to remind themselves. Nature itself was their Sphinx.

"The British were going to hang me," the Chief said one day, "I had killed a man because he had done something bad, and as the Chief's son, it was my duty. The British said that he should have been tried by a court to determine his guilt and his punishment. But both his guilt and his punishment we already knew. And we could not be certain that the British courts would deal correctly with the action of this man. The British courts have been known to allow the guilty to escape. So the British took me to prison. They sent me to court. They convicted me of murder and sentenced me to hang. On the day the British took me out to the gallows beneath a big tree on which they planned to kill me, they found themselves surrounded by my father and thousands of Dinka. The British did not like to give up when they had made up their minds, but if they killed me, they could see that they would die. Finally they let me go. My father and the Dinka had just stood there, waiting."

The boat train stopped against the bank at Malakal. Small official buildings and thatched huts. World War I had nearly started here, years earlier than it did, between the French racing for an East-West African empire and the British racing North-South, Cape to Cairo. That time line lay in wait for its chance to live, opened fire from the French ships upon the British in 1942 in Morocco. A corpse waited for the boat train of the week ahead of us to return and take him the first step back to his homeland. A European. "He took malaria pills and then sat in the sun with his shirt off all day. He didn't know the pills weakened you against the sun even as they protected from the malaria."

The Chief led six other Dinkas and me down a trail into the bush outside the low buildings and huts of Malakal. In a hut, the old woman of power poured us drinks, and we smoked the cigars of concentration.

"We Dinkas are the bravest in all Africa. We must kill a lion to make manhood. Unlike the Masai who first circle the lion before the man makes the kill we must go out alone to do this. The other thing I did to make myself a man was to go up alone into Ethiopia and smoke opium so that I could last all night and make love to Galla girls, the most beautiful of all women. Then I came back to be the son of my father and succeed him as Chief."

"In between I helped capture two Italian generals in the war as a Scout, and I worked for Pan-American in Khartoum."

"And you go back to the tribe after all of that?"

"I have never left the tribe. I go out to see and to know the world in order to help my people."

The boat whistle blasted through the bush, at least three kilometers away. I made a motion to get up and run back. "We have four hours. Big warnings here. Let's finish the meat."

One man lay sleeping, passed out from too much drink. "So we have to carry him back?" Laughter. "No. He can sleep for ten days till the next boat comes." "That's a long time." "Next time he doesn't do this." Laughing, well-fed, we strolled back to the ship, the Dinkas folding over in merriment at the surprise awaiting their fallen comrade upon awakening.

After Malakal the Sudd grew thick and the boat sometimes could barely make its way through the wide placid waters, almost a swamp. Hippopotamuses, crocodiles, birds, fish.

"If two Dinka go out and kill a hippopotamus, one must stay, standing on the hippopotamus and with his spear keep off the crocodiles until the other returns with more people to take the meat back."

Dawn. Heat. Buildup of equatorial clouds to downpouring fresh cleansing rain. Sunset. Lightly cooked dough. Night. Sleep. My life rebuilt itself in dreams, rest, watching, listening, sharing alertness.

A burned village on the left bank. The Nubas kept their rifles at the ready. The Arabs kept a closer look from the all-but-empty first and second class steamer. "They want to turn our world into Islam and agriculture. We fight, but they have airplanes and many guns. A hundred villages like that one. The British whom we ultimately trusted betrayed us, leaving us under the Arabs, not giving us our own country."

The next day the Shilluks and Nuer began their songs, dances, and chants. Their villages were being approached on the right bank. The women broke out of their segregated area to join them. They were reissued their spears. They averaged over six and a half feet tall, almost naked, without a trace of fat, wiry where the Dinka, a little shorter, were thickly muscular in the chest and shoulders. Their radiance, their exulting over their return to the tribal homeland washed over me, until my body began to move with the same waves of joy.

Behind the forms and in the precincts of the Sphinx, the stepped pyramid, the sacred bull, a small group of esoterics carefully hid their searches through the landscapes of ecstasy from the eye of the oppressed, depressed, distracted, and power-mad, but here in the south still remained the ecstatic unity of an entire people, armed, self-regulating, with herds of cattle. A living part of the living cosmic landscape. More joy here, in this village flocking to meet the boat train, in this reunion than in all of Europe, all of America. More than joy: jubilation.

The Dinkas I stood among watched as intently as I did. Now they were close. Juba was the next stop. "But it will not be so happy there," the Chief said. "The Arab army has its headquarters there and they watch us closely. But it will still be happy."

Afterwards, after the jubilating at the dock of Juba, he let me walk with him to the Dinka village. Many huts. Lost in the maze, except for following him. A door opened in the staked wall around a large thatched hut. Two women greeted him, falling to their knees. More people appeared. He sat down in a great chair in the middle of the hut.

"I cannot ask you to remain more than one night. We have so much to do."

I understood although I would have carried his food to be permitted to stay. My very presence could constitute a danger. Khartoum wanted no one in the great world to know of its deadly war upon the peoples of the South. The worldwide war of the jealous and greedy upon the well-kept sex and lands of the few tribes that still ruled their territory. I sat on the bare ground by the side of his chair as he received and talked to people.

From this Chief's presence, intelligence, and personal prowess to priest, king, president, dictator, and general secretary. Downhill all the way. I had reached the point before Egypt, even before the

Atlantean league of tribes, the Solutreans, had created their megalithic legends. I had finally met a real man, almost the rarest of beings upon the contemporary planet. A man who could travel, speak, fight, lead, jubilate, love, decide, keep silent, listen, be. A higher aim than being president had at last arisen and I could see very well why Jefferson had not wanted that position engraved on his tombstone. Conscious man had not yet even reached my ears as rumor.

Bones and Masks of the Ancestors

YEARS PREVIOUSLY ON an archaeological dig of grass over Indian mounds by a creek in the northeastern Oklahoma hills, with a spatula, a camel's hairbrush, and acetone I had labored all one day to rescue five two hundred year old wet bones to be recorded as one of the small bits in the great effort to reconstruct man's prewritten past. At that time, at nineteen, the thought of five hundred year old mounds and a vanished civilization filled me with excitement, half a millenium seemed so ancient.

Now my interest reached much farther than that. Back behind the civilizations, back into the tribes, I had finally found a man who satisfied all my imaginings about that name. Who understood nature, women, war, injustice, modernity, resistance, silence, joking, tempo. Who had been benevolent, courageous, intelligent with the gift of empathy. I began to see the glimmerings of what I wanted, and for the first time since Tangier I knew that Beni Sidi could not get it for me, nor could anyone help me that I had ever met, not even the Dinka himself, for he had been created by his tribe, his position in that tribe, and by a certain moment in history. I wanted to know objective history for the first time. Where did man come from? Upon what had the tribe been built? Where might man be able to go? Where might I be able to go?

Reaching Nairobi I devoured books: Leakey, Lorenz. At night energy from drinking cheap beer and inhaling the excess sexual energy floating off the cheap beautiful perfumed girls. Books about the on-going discovery of the genus of man, species ago, millions of years ago, in the hot dry Pliocene savannah.

Wolf-packed, primate-brained, tool-using, promethean theft of fire from lightning. Out in the open, not hidden in the trees. My brain afire, my mind excited almost beyond return, I headed out hitchhiking for Tanganyika. Mountains ablaze with fires at night. Drums. I felt so attuned that one night I lay down to sleep in the open, certain nothing could happen. Two men appeared out of the dark. "No! No!" They took me into a crowded hut and let me sleep in a corner. "Hyena. He will come and bite off your face. Jaws so powerful!"

Remnants of the birth landscape of the genus. Atomics, Space, Jets, Empires, Civilizations, Cities, Tribes, Magic vectored from this upright genetic force coalesced with tools and the big brain on these savannah environments. Before us, before Neanderthal, before Erectus.... I sat for days and stared into the dusty trees, into the magnetic African sky until it crackled purple light.

One day, lying sprawled, exhausted on a tiny wooden footbridge outside a village someone walked by, but, my thoughts finally cohering, I refused to open my eyes. He tenderly shook a foot-long piece of sugar cane in my mouth. I kept lying there, eyes closed, thoughts cohering, applying tooth pressure to the cane, tasting the wet sweetness dribbling in my mouth.

Then on to Zambia, newly granted internal independence, but where the British retained management of external relations for another year. The Union Jack flew in this clear savannah woodland air. Last moments of the most extensive empire in world history, ironically in the billions-year-old plateau heart of Africa, far from the coast. Extensive copper deposits; my mining engineer debris automatically whispered.

The young British officer asked me to tea. The only European in ten thousand square miles except for the occasional medical team, missionary, or inspector. Spotless uniform, behavior correct, friendly to my purpose of wandering over the landscape of origins.

"I had over three hundred dead of smallpox in the last two weeks," he said, "but I cannot get vaccinations made. There's a witch operating here who says the needles will sterilize them, a British plot to reduce their population. They go to her for healing. I pick up the bodies in my truck." The rush of the tropical sunset, storming silently over our teacups. The right side of his upper lip flickered a derisory understanding. He gave me a letter to the friend of his commanding an army post near the border of the former Belgian Congo, now in a civil war determining its new future. One of its Maoist Chinese-backed armies was heading in the direction of Zambia from Stanleyville.

At the small outpost, one hundred and fifty miles away through the savannah bush, no horses or cattle due to the tsetse fly, which I kept looking for lest it bite me and deliver the sleeping sickness, the first morning that I wake up, the lieutenant, Jim, said, "Sorry, you'll have to move on today. We've just received orders to advance to the frontier and halt the army coming from Stanleyville should they try to cross." He'd just shot two cobras in the tiny rose garden outside my sleeping room.

Off we went in two Landrovers, one pulling a trailer with a machine gun. Two British officers, two Zambian sergeants, and twelve Zambian soldiers. The air seemed giddy with glory and insouciance. To stop an army.

When they let me off to continue on my way they turned off to go to their assigned place on the border. I stood, watching, unable to move until not only the sixteen disappeared, but their dust had settled.

That's how the feeling was, I thought. That's how we came out onto the savannahs from the trees, that's how we faced dangers, continual, and from all sides, hardly conceivable today. That's the genetic drive underlying us all, masked as it might be by the conditioning of today's social megamachine. Insouciant, giddy, certain somehow that we'll do it no matter if the odds are ten thousand to sixteen.

For another three weeks I moved through the level African uplands, going back into the mammalian Pliocene before earliest man arrived. The spray lifted up like an upside-down rain from the crashing waterfalls of the Zambezi, forming a small local rainforest. Back upriver from the falls I walked, away from anyone. A group of elephants, each trunk holding on to the tail of the one in front except for the leader, an old bull, tuskily picking his away across the river. Then, after I turned back into the bush, suddenly a herd of great twisted horns, the giant Kudu staring at me, I staring at them. Becoming a Kudu, the magic of Kudu changing my physiology, I became humanoid, moving fearfully- alertly through "the bush", tranced out, and saw through joining the Kudu in their stately movement, the vegetative Kudu world, their flesh alert to the ever-possible predator, how totem came about, recognition and absorption of the power of a species. How the patterning of the organized sensitivity of the higher mammals pulled into the memic circuits of culture to dance a synergy with the genic circuits, an objective art blowing the synapses of practicality, producing an audience hooked on the oral-visual transmission, the theater of transformations.

Life, evolution, powers creating totem, art, ecstasy, alignment, and we humans only the latest and not the last on-flowing of the great effort. The African night growled and chattered starry grandeur through the acacia columns. Waking, the red dawn chill and my hot heart about to burst.

At last, knowing how I had gotten here, ready for where I am going. India. Retrace steps to Mombasa, sail to India. The past could be sought and found through its traces, absorbed by the shock of entranced recognition. But the future could be seen only by enlightening the darkened mirror. Now could begin the search for the mind that spawned intellect and ego.

Being-Confrontation

WAITING FOR THE British India ship's departure from Mombasa for Bombay, I lived with Kumar, a slim young Brahmin, and his family in Nairobi, preparing for India. Being unclean, I could not eat with his family as this would have horrified his mother and grandmother. But we drank cups of tea with boiled milk in the shade of the Banyan tree in their small courtyard surrounded by one story rooms in one of which I slept on a brown twine string bed.

The battle of Kurukshetra filled my mind with its roaring elephants, thundering voices, clashing of armor, and the poetry of Krishna revealing the magnitude and density of the danger and glory of the Cosmos besides which even war on a continental scale fell into its proportionate place below the necessity of a man fulfilling his destiny if he wished to transform and escape being forever meat for the butchering machine. Or as Shakespeare put it a brave man dies but once, a coward dies many times.

Kumar's sister-in-law delivered a child which died two days later. His mother wrapped up the corpse and gave it to Kumar. Taking a shovel together with the wrapped baby he and I walked to a burying ground. He scooped a shallow grave, deposited the wrapped baby, and covered it. "We have no ceremonies for young babies." he said. "My sister-in-law ate too much betel. She did not take care of herself." I watched a mass of useless American sentimentality towards babies and cruelty toward young women shrivel in the dry clear Nairobi air.

Kumar had travelled with little money through many countries of Africa by himself despite a definite anti-Indian mood in several of the newly independent nations. All Brahmins are taught to cook to insure purity, but also as he said, "this means we can also support ourselves wherever we go." His father had taught him to repair radios in addition so that he could cope with the modern age. It appeared that he studied me to be able to deal with America and had offered me a study of India in return. Correctness marked our every move for two weeks. "Goodbye," he said as I left for the ship after two weeks. "Goodbye," I said. It was clear there would not be, perhaps could not be, any further contact. I watched another American shibboleth wriggle into disappearance, the belief that an existential future was necessary to make a friendship meaningful.

Dulu, the Wakamba girl, so beautiful, sat beside me in the bar by the docks. The bar full of romantic music, "Long Ago and Far Away," "Begin the Beguine," "night of Tropical Splendor," beer, sex, and ganja smoke. "I know what you are saying. You don't have much money but you wish to travel throughout the world and find something. You remind me of my boyfriend. He always travelled by himself. A lion killed him one night on his way from his village to mine. I love you. I can't sleep with you because you don't have money and our mother would not like that, I have to sleep with a sailor, but you can come to our place with me and sleep and eat until you leave."

I sat in the front of the cab with her and the Danish sailor, also beautiful, but blonde, in the back. We went up a stairway to the rooftop of a good sized building. Rooms were all about the roof. "You sleep here." She pushed me into a small space with a narrow string cot.

The deep sleep of the warm African night and of a satisfied unconscious fell instantly upon me. In the early morning I woke, went out onto the moonlight flooded whitewashed roof, the rooms' rusty iron roofs glamorous with refulgence, and walked silently over to the room I had seen her go in. Through the small air window I could see her and the Dane sleeping peacefully, chocolate and cream, both turned golden headed as Douanier Rousseau could ever have imagined. In the morning a young girl brought me bread and a bowl of stew. For three nights, I fell asleep and healed in the warm glow of this girl's love who somehow had secured me this place to sleep and eat. The day the ship left she said, "I love you. Get some money for us to live on and come back to me after you have found what you want."

The great blue musical waves topping the great blue mute swells of the monsoonal westerly helped carry our ship along the northerly of the Sinbadian routes, the great orc now extinct on Madagascar to the south. Four hundred Indians, leaving Africa, debris of the end of the British Empire, fearing for their lives and property under the new Black nations, lay side by side on the floor beside me in steerage. Each of us had a space about six feet by two feet. Barefoot, uniformed Malays served us noiselessly our cheap adequate food, vegetarian, since nearly all the passengers in our class were Gujarati. A Scottish engineer made the rounds, an English officer looked in on things. The ship still imperial with its impersonal analysis and synthesis of ethnic efficiencies.

One other paleface in our lot. He never spoke. "He's going to visit the guru who never speaks," said one Gujarati to me. "Already he has taken the vow of silence."

The Gujaratis conversed to me only of economic losses on leaving Kenya, how much they could take out with them, what they could do with that amount in India, the relentless overt mathematics of commerce of the crowded subcontinent that was the religious duty of their caste.

The ship stopped. I awoke with a start. The Gujaratis were already packed and filing for the railings. India! But at the rail I could see only a long low line far in the dawn distance. Wooden dhow-like lighters surrounded the ship.

"The Rann of Kutch!"

Slowly through the morning the Gujaratis disappeared toward that low featureless salty desert's shore. Tens of thousands had gone to Africa, perhaps three, four times as many washed back after two generations. Not one outburst of open anger or tears had I seen all the way from Mombasa. Perfect order had obtained throughout the voyage. No Americans would have been capable or would have desired to be capable of such a feat. Perhaps no Western society. I pondered, watching a culture order such harmony out of the chaotic bitter occurrence of exile. These were the people and forms of action among which I hoped to find a way to reach the state of consciousness and the level of being necessary to see the future for, without knowing why and what, I had made up my mind to do no other work but to seek until I found or was found.

In Bombay I stayed with African students, still lingering at leaving the last vibrations of the worldly educated heart before entering the source-palace of Aryan thought, sublime and austere.

One took me to a "good house." The six girls, of striking diverse types, eyed me easily, sitting in the living room with the Mother. An attractive martial-jovial voluptuary made advances. The old habit-body got up to go passively with the lady, but the new chemistry surged against this and I watched myself catching up the one, not my sexual type at all, but the very embodiment of the pictures of Sita, the "ideal" Hindu woman, fully and harmoniously developed. I wanted to learn from her.

You must know something of the women before embarking unafraid in this country, I told myself. I had come looking for this type. "I am poor," I said when she asked me for a little more, "but I am paying you the rupees we agreed. I must use all the rest to find the road to Badrinath."

Beautiful, she made love perfectly.

"You made love so well."

"If you had not told me you were poor and saving the rest of your money to seek, I would have made love so that you could not resist returning to me until you had nothing left."

The African told me, "I did not have the money to go in with a girl myself, but it did me good just to sit there while you did."

The hot May pre-monsoon heat felt like hands on my back and head pushing me down, but I walked like a wound-up machine down the dusty road from the Banyan tree shrine and its shaded rock sitting place in the center of the village by the Ajanta caves. At the caves, my machine came to full stop. Yes, imagine girls such as the one in Bombay with their full bodies and fuller sensations refined and exalted in schools of the dance of cosmic ideas, at the disciplined height of ecstatic trance, beheld in the eye of a similarly transformed painter-meditator, himself organically transfixed in the simultaneity of a flood of sensation, the sun of the world of forms, floating on the ocean of non-perception. The marvelous paintings of girls showed that this had happened here at Ajanta. Liberation had meant freedom for sublime beauty to manifest completely, not hatred of life and love giving freedom for the killers of beauty to ravage the girls and stunt the boys into stuttering manifestations.

Now I realized that after the lights had gone out for the West with the fall of Rome as a result of the millenia-old Chandala rebellions of the Middle East and Europe which then had inflicted the guilt cult's incubus upon the twisted wreckage of ancient teachings, that real knowledge had continued in unabated power upon the subcontinent for centuries, and perhaps was not yet totally extinguished. A delirium flushed my already overheated head and pounded my heart.

"Who do you think could have created these perfect lines, this extraordinary landscape?" A voice sounded within.

"By the same kind of man who wrote the *Mahabharata*. There were more than one and at least one probably still exists. Now I know what I'm looking for." I left Ajanta immediately without going to Ellora, took a Tonga to the highway and immediately began hitchhiking to Old Delhi.

Indore. A youngish man on a motorcycle who had driven by me twice picked me up to take me for tea at his bungalow house. "Look at my servant closely," he said. I had been told to look at animals closely, at plants, at things, but never at a human being. Always at a distance and covertly, as if approaching an incipient taboo, or a direct danger.

"No, closely," he said. "Examine him closely. What race is he?"

"Not Negroid, not Caucasoid, not Mongoloid."

"Yes, he's different, he's from the old race that ruled India before we came. We think our yogi came from these fellows. Now, they have only magic left."

The Aborigine seemed to understand all that occurred and stood stockstill, under command but in no way diminished in being. Understanding passed between us. The master examined both of us closely. I could not examine the master of the house back. My eyes could not roam over his appearance at will, although he made no signal to stop a reciprocal examination. I experienced a tremor of nausea, seeing that I was less than he. I realized that I was and had always been afraid of man. The knowledge incarnate in my host did not stand in fear of examining man.

"I must take you back to the highway. My wife returns shortly, and it is impossible for you to be here at that time."

A two-and-a-half ton truck driven by two Sikhs stopped and I climbed in the back upon their load of I-beams. After an hour we stopped at a small shrine surrounded by roses. The driver plucked a rose and ate it, petal by petal. He handed me one and I ate it. In the cab, driving away, he said, "I drive a truck, an American drives a truck. He makes five-ten times what I make. Why is this?"

"Well, he can't stop to eat roses, his highway is smooth and wide, he drives twice as fast, his truck is eight times bigger, and he usually doesn't have a fellow-driver." I thought but didn't say. I said " Who can say?" And happily watched my Mr. Know-It-All, who always had to give a full explanation based on partial knowledge, shrink a little more.

The truck stopped again at a big truck stop, dozens of small booths, twine-string beds, and a small shower-house. The Sikhs in the shower house turned on me, glaring and yelling, when I tried to strip to shower. I had to keep on my khakis.

Outside, the sun poured down glittery weight; I could sense its pressure exerted continuously from ninety million miles away. Halting on my way to the tea tables, I knew I would not move another step without at least an interim answer. A pile of broken rocks, about five feet high, stood at my side. I crawled up and lay on my back, arched almost into an upside down "V" by the cone of the pile. All the tenseness in my body left. The sun beat down. I sweated. The Sikhs continued their truckstopping. Why? Why? Why?

A Hindu word formed deep inside. Images: worlds appearing, disappearing, individual lives, species lives, biospheres.

"Lilas." "Lilas."

Delight. Brahma continues the coming-and-going of universes because of delight.

Delight.

The rocks have become as soft as pillows. The sunlight thickly golden. I breathe the thick gold heavy air in and breathe myself out. I have to struggle to imprint a trace on my organism, to make a small profit from the intoxicating prana.

The truckdrivers stand over me, brown-eyed, impassive, knowing. "We go on to Delhi," one of them extended his hand to help me off.

The Sikh let me out in the early morning that late dry May by the banks of the Jumna River back of the Red Fort. Coming around Red Fort, I knew this street was it. "Chandni Chowk," a man told me. I plunged into its restless life, not turning back once to look at the walls surrounding the ruins of Mogul warlike glory.

Seeing a Sikh Gurdwara on the left, I tried to secure a place to sleep.

"No foreigners allowed here. Go to another Gurdwara."

Suddenly hopelessness sucked up the energy glowing in me. Hundreds of people streamed by every minute. I crossed Chandni Chowk mechanically, and continued up the right hand side. I gave up. I might know why but I had no more idea of what or how than ever. No place to sleep. The faces that flared up seemed no more awake than in New York.

At that precise moment that I gave up, a young Hindu woman stopped me. "Can I help you?" I came to a complete stop. She had committed an utterly taboo act. "Yes. I'm looking for a place to sleep."

"Come with me."

She led me about two blocks further on and we turned a corner to descend into a large basement room filled with bearded men in yellow clothing, sitting, standing, walking, lying down.

"He cannot come here. He cannot come here. Take him out."

"He is poor. He needs a place to sleep. He's seeking."

"He must go. Now. Quickly." Two of the men advanced toward us, gesticulating.

"Don't worry," she said outside. "I knew they would do that. I wanted to see if you would live with them. I knew these fake holy men wouldn't live with you. Follow me."

We turned to the right to avoid Fateh Puri Square and then into a side street. A stair led through a small door up from the street. A man appearing like a beggar quickly jumped up and opened the door for her. Ascending two flights of narrow dark stairs, she opened another door.

Asia! At last I had arrived. The spacious courtyard, blue-tiled, with a fountain in the center, shone cool in the sunshine. The courtyard was surrounded on four sides by two stories of single rooms which overlooked it.

She brought me into the large room on the right of the courtyard. A man sat in a chair, legs easily drawn under him in spotlessly clean kadi soft cotton shirt, and wearing a dhoti pulled between his legs.

"He will say if you can stay here."

The man looked frankly and interestedly toward me. I was suddenly afraid that in my beard, dark sunburned face, spotted khakis, cheap plastic sandals, and with my small Moroccan straw bag over my shoulder my external impression would turn him against me. I refused to become hopeless. I looked back at him, compact, confident, powerful in an organic, relaxed way.

"American?"

"Yes."

"Oh, I thought you were a Russian," she exclaimed. "Our Prime Minister, Nehru, has just died, and one of the last things he said was to be friendly to our Russian friends, and so ..."

"And how did you arrive here?"

"I left my businesses and my writing in New York and have crossed North Africa, journeyed up the Nile and to the Zambezi, taken a ship to India, and came by truck to here. I am searching to see how other people live and to find what I must do."

A man and a woman, in their forties, now joined us, about fifteen years younger than my judge.

"This man," he summed up turning to them, "studied to become President of the United States. He is highly trained but he realizes he does not know something he needs to know." He turned back to me, sitting stunned stonestill by his piercing my disguise. " I will make a deal with you. I wish to know how the mind of a real American works. You will put your mind completely in my hands, and I will show you—India."

Having by now lost all interest in the contents of my mind, which I could see were worthless to anyone but anthropologists or politicians, without hesitation, I said, "Fine."

"My method," he said, "consists in making a being-confrontation for four hours each day with someone I have never met before and about whom I have heard much concerning his power. We will accomplish both purposes, my seeing how the American mind works and you seeing India, by you accompanying me to these meetings where I can observe your reactions. The only rule: you must answer any question I ask you, and do what I request. Otherwise remain silent and observe what you can observe."

I had found the man I had so intensely sought. I could detect no opposition whatsoever in my body. "Alright."

"Now," he said, "Usha will show you your room, and you must pay her some attention so that she will be happy with your staying here. These are her parents and you must do what they tell you. I am a friend of the family staying here for awhile."

The next dawn I rose from my twine-cot on the roof to see Usha's two brothers flying kites high over the packed city below. "Every day we fly the kites."

I went below to the courtyard and greeted Usha. "Good morning! I slept so well thanks to your hospitality. You see, I'm in plenty of time for breakfast."

"No, you haven't bathed yet! You must immediately bathe yourself. We must bathe every morning."

"We bathe at night."

"Here, you must bathe every morning." Laughing, but stern, she showed me to the bathroom, where a dipper stood by a faucet protruding from the wall.

I poured ladle after ladlefull over my head, soaped down, and poured some more, rubbing my head and skin briskly till all the soft dirt and soot and sweat from my trip had scoured out.

After breakfast, Anil took me on a slow walk down Chandni Chowk with the grandfather of Usha, stepping into a clean cool shop back off the street to insist on my buying a light finely woven kadi cotton shirt.

Then we walked on to a bathing club next to the Jumna River. The grandfather lay down on the grass, after removing his garments. A masseur worked him over thoroughly, oiling, rubbing, kneading. " He does this every day," said Anil, "but I follow the old tradition in that I come here to swim only on the night of the full moon and of the new moon."

Anil made me take off my clothes, and he and I swam slowly about a hundred yards down and back up the Jumna.

" Don't worry," he said. "You are afraid to get sick, but this water will not make you sick. Everyone calls it a holy river. My son says it's because the fine clay particles in it act like one of your zeolite sewage cleaning plants in America."

When we walked back to Chandni Chowk, the grandfather left to do his business, while Anil ordered two mango milkshakes for us.

"Well," I said, "I'm ready for the being-confrontation. Just tell me when."

"You just had it for today," Anil said, pleasantly insouciant. "I had it with you and watched how you hated to be with an old man, how you didn't like to change your clothing, how scared you were to swim in the river I swim in, and how nervous you were about drinking anything with uncooked milk in this restaurant. Nonetheless, you did it all, so do what you wish with the rest of the day, and tomorrow we'll go somewhere else."

The next day, arriving with him at a large Delhi office off Connaught Circus, I followed Anil as invisibly as I could into a spacious back room where sat a group of nine focussed men. The air sharp with tension, expectations, paranoia. The slim hawklike central figure, a member of parliament, controlled one of the large newspapers of India.

"You are a powerful man," Anil said after introductions and a moment of silence in which they had looked at each other, "you are a member of Parliament, and have control of our greatest newspaper, but you are full of pain that you dare not tell anyone about. You worry for your life."

"True," the man said "Exactly true." He, like the other men, was elegantly clad in casual western clothing including light sports coat. Anil in dhoti, his powerful legs bare, and khaki shirt. After another short silence, the man lowered his gaze and said, "Help me. Help me if you can."

"You must submit to me. You must allow me to do exactly what I must and tell these men not to stop me no matter what it seems to them I am doing."

The talk changed to Hindi. I watched the physical scene move forward, while taking in the emotional sounds of the words and the emanations from the solar plexuses.

After the man instructed his staff, Anil, of deep and broad chest though of short stature, stepped quickly forward, seized the man and laid him flat on the big desk. His hands began to work on the man's body vigorously, especially in the region of the belly. Some questions from Anil, some answers, gradually the man grew limp, his eyes remained open, conscious of his body but unreactive to the environment. Anil had removed all the man's clothes except his underwear. The staff stood in unmoving silence, eyes intense.

"Feel him," Anil commanded me holding forward one of the man's legs. I had never handled a body like a thing, and had never even conceived or fantasized doing such to a man of power in a modern office surrounded by an obviously loyal staff of managers.

This was far beyond looking closely at a man. I shivered. But I had promised to do whatever told.

I picked up the leg beneath the knee and held it gingerly. "Feel it, feel the thigh, move the muscles around, move them however you wish, see what they will do. You will see, he's perfectly loose, and has no reaction."

I kneaded the thin thigh muscles. Not a man moved in the room though I knew this violated many of their taboos. Sweat prickled my face and hands. I experienced to the full my fear of man and had to force my hands to keep moving through an atmosphere I felt to be more resistant than the man's muscles.

"Give me back the leg."

Anil sat on the desk half facing us, half facing the man, and continued to talk to him in Hindi while working his body. The man's eyes remained open and unblinking.

After what seemed an immense time and yet very quick, and indeed the whole scene lasted nearly the four hours Anil had spoken of, Anil got up and stood facing the man who came to, stood up and put on his clothes.

"You have cured me," the man said, "and furthermore what you told me has cleared up for me what I must do at this stage of my life. What can I give you? I will pay you whatever you ask."

"I do not want money. But in the future if I ask you to do something, please do not refuse me."

"Of course. I see that indeed you are everything that has been said."

"And do not talk further of this. It shall remain between you and me and these few witnesses."

"You see," Anil told me as we walked back to Chandni Chowk where he ordered two mango milkshakes, "the object of these being-confrontations is that I or he shall submit, whichever has the most being."

"But in America not one of us would submit to another, at least consciously, we are free."

"We do not submit to anyone's force anymore than you do, perhaps less. I speak of submitting to a higher being. I should be happy to submit should I find one, and in the past I did submit to two men, now both dead. My power in India consists in the people who have submitted to me and I do not ask them for anything even though I tell them I might ask so that they will not feel in debt to me, but they find ways on their own to help me and that kind of help is useful."

The following "autobiography" did not emerge all at once, but is put together from a number of scattered sentences that Anil uttered during the next three weeks. I never forgot them.

"When I was six, a bearded man stopped my mother. 'You do not know how to deal with this boy,' he said to her. You must give him to me. My mother struggled within herself for only a short while. 'Very well, she said, you can take him.'"

"I lived with my master until I was sixteen in the Himalaya mountains of Uttar Pradesh. In that ten years, he showed how to do so many things, to breathe, to speak, to cure, to see, to understand all kinds of people ... I would never have left except that he died."

"Going to formal school the next two years at one of the hill cities, Dehra Dun, and then eighteen of my classmates came to me on my eighteenth birthday. 'We have sworn to follow you the rest of our lives.' Not one has ever gone back on his word. I can visit eighteen cities in India and act there as in my own home, just as you see here in Fatah Puri, because he was one of the eighteen."

"At eighteen I heard of Gandhi. I traveled far to see him. I said, 'My caste forbids me to bow my head to any man even if he shall strike it off for my refusal, but I am here to serve you without bowing my head.'"

"I became Gandhi's bodyguard and in thick crowds he sat on my shoulders, his legs around my unbowing head, and I took him safely everywhere."

"Gandhi became my second master. He did not know as much as my first master, but what he did know applied to a different world from the individual, the world of the public and history, and I learned his teaching very well. His teaching was not saintly pretense, nor anything that can be imitated, his teaching was action, concentrated, one-minded actions. Not many learned this from him. We that did called Nehru a beautiful boy."

"After the first two years, sometimes I would cover as much as two hundred miles a day off the main roads on relays of elephants carrying Gandhi's message, speaking to every kind of village, every caste, debating the program, organizing the action."

"After Independence, I was assigned to Patel. The people loved the beautiful boy, so he must govern, but we must save India. Hyderabad verged on joining Pakistan. The West, the East were gone, the Center seemed ready to secede. I organized trains full of men from Bombay. We arrived in Hyderabad, thousands of us, surrounded the ones calling for succession, surprised all of the hundred of them at their meeting, cut off their noses so that no one would ever forget, and took power."

"Shortly, India experienced a great coal shortage. I became head of a two thousand man mine and we showed how to increase production. I had never mined coal before but I knew how to produce the action that mined the coal. When India was in critical need, at that point I went to work, and when the need left, I left at that point. Only a few thousand people know me, my name, and my actions. None know my life. That has been a great secret to my achievements. Credit always given to others."

He took me to an outcast village. "We don't call them outcasts any longer. We say Scheduled Castes." He grinned sardonically. They lived in concrete boxes, and worked for the village farmers, the Jats. Anil launched into a fiery speech with many emphatic gestures. The men crowded around him, listening, silent, staring, burning eyes. Anil wound up laughing, embracing some of the men.

"I told them, are you men or not? You must organize your lives, make sure of your water, fight for your rights. India is for all of us."

"Have you ever been here before?"

"Their leader met me and asked me to come here to speak."

One day he said, "I'm taking you to a famous Guru. We will see how real he is."

We stopped to buy some bananas. "Always take some gift, fruit is best, flowers are good but not eatable, when you visit these fellows."

"You are supposed to be able to see into the other world," Anil commenced after the preliminaries had relaxed everyone. "This American's mother died last week. Where did she go and why?" He looked at me. "You can tell me how accurate he is," he said. The Guru sat up on his couch, put his head back, rolled his eyes behind his eyelids, his hands resting calmly on his knees. After a few minutes of silence, he spoke. "Your mother was a good woman, but she had a limited vision of the cosmos. She was very interested in food, in money, in things. She worked very hard but she could not see in that lifetime beyond this world. She went to the fourth heaven, the middle heaven, where she is looking down, ready for rebirth. She believed in God but she never attained the truth and reached happiness. She has much understanding left to gain."

Tears went down my face. What an exact analysis. I had sat in the Embassy in front of the consul only a week ago as he gave me the telegram. I could not say a word, could not stand up. The tears had flowed automatically. I could detect no emotion, no bodily movement, but the tears had flowed as if by their own will. The consul had sat and watched me, not unkindly. When I got up to leave I knew that at last I was free. My kadi shirt clung to my chest, wet from the tears. Something had washed out from the innermost chemistry. My mother had blessed my search. And then she had left my life and as she did so somehow set it free.

These tears were different. I felt the immense pain that informed ignorant life, that sculpted the confined contours of average existence. Of the utterly stupid white bread of America, utterly because no existential requirement of poverty, of tyranny, of caste required it. A culture that had mastered matter, technics, politics, and forgotten poetry, sacrifice, and glory, along with the taste of good bread. Of how my mother had wasted her potentiality.

"I'm sorry," the Guru said, "but I must tell the truth as I receive it, or my power would leave me."

"Of course," I said.

Anil said to the Guru, "you have a pain in your upper belly for years, and you have told no one, but you think about it every day."

The Guru fell to his knees in front of Anil. "If you can see it, you can heal me. Please, kindly do so." And Anil did so.

After three weeks of these daily confrontations, Anil told me he had to leave for Gujarat. "But first I will take you to a Hindu wedding."

Color. Crowds of People. The man and woman sitting formally, costumed, upon the stage. Hours. Talking to the Professor of Sanskrit at Delhi University: "Anil Thakkar introduced you to me, he said, and so I will talk to you, because perhaps for once in your life you have a chance, though undoubtedly not easy for you to take. I call the fearless man God. Anil Thakkar fears nothing, man, woman, beast, god, disease, old age, death, politics. I call Anil Thakkar God."

I would never have dared myself to say that. I knew it was true the moment I heard it and I realized that my cowardice extended not only to all the forces that Shankar had just enumerated, but even to the world of ideas. That man could become God I could not have thought, much less have said of anyone. I saw that my entire existence was based on fear, and suddenly another fear took possession of me, the fear of fear and I knew that at last something had come that would permanently recur, different from all that had been before.

My entire life came to an end in the color, the heat, the crowd, the blared wedding horns. I had met God and I judged myself and saw nothing. I did not wish to go forward with anything of the past. Every idea, emotion, sensation, movement, action, thing, situation, search, finding that had ever passed through my organism had been rotten with cowardice. Even my experience of cosmic delight upon the jagged rockpile had been only by permission of Brahma.

I woke up. And then I died.

Shankar watched me. He could see what was happening till I died and then I saw he could see no further. I had disappeared not only from America, Egypt, and Africa, but from India. I was in nowhere and had stayed awake. Time, "that must have a stop," did stop. The movie kept playing in the lower right hand room with all the fabulous techniques mastered in Tangier. I had no idea what to do in this new world. All possibilities became clear to me, and my utter powerlessness to determine which route to take to master these possibilities and to select wisely from among them also became clear. I could sense my body straightening, lifting itself strata by cellular strata. I would ask Anil to take me to that first step of new life before he left me for his trip to Gujarat.

"I must leave Delhi for several months," Anil said. "You cannot go with me where I must go, and there is no need for this. I know where you want to go, and I will ask another of the eighteen to take you."

A lean, mid-fiftyish student of Anil's, a pharmacist from Allahabad, took me with him, at Anil's request, on a journey to retrace the steps of Yudisthara, the last survivor of the wars of the *Mahabharata*, on his way to die consciously at Badrinath in the Himalayas.

Samir and I descended from the train at Hardwar, Hari Dwar, the door to God. The Gunga pours through a rocky gap at Hardwar to debauch upon the northern plains of the subcontinent. So Hardwar is the "door" to the Himalayas, who must then contain the original image of "God" to the Hindu mind — elevation? snow/ice? hermitages? hardships, adventure, and beauty? the direct tradition and experience of living men who have attained? A slim, exquisitely muscled yogi advanced toward us over the blue and white tiled floor of the small cool temple fronting on the rock rolling river.

Learning that I was from America, he became quite animated. "Please, sir, I must go to your country. There I will be appreciated. I can show your countrymen such potentialities!"

He began to do his series of exercises. Samir and I watched him. No one else appeared in the little temple. The pounding of the Gunga rose to a roar in my ears. Here was I, searching for elusive reality, traveling in the *Mahabharata*, identified in my mind with Yudisthara, certain that in India I would find the missing knowledge, and here he did his Cobras, Lions, Plowshares, certain that only in the U.S.A. could he achieve his destiny.

His dark eyes glowed into mine. I admired his body, his discipline. I had done just enough yoga to see that he had mastered his hatha. But my interest lay at this moment totally in the path of raja yoga, the exercises of mind mastery with the object of permanently breaking through to whatever lay beyond the field of thought. From the peyote experiences I could see exactly how far this yogi had gone and he had not broken beyond the field of breath and tissue. However, he expertly and comprehensively probed the boundaries.

I watched my fantasies arrive and disappear; first, of platonic love, then of becoming an American chela and studying his system, then of blowing his mind by spiritual denunciation, then confessing my state of inner desperation by falling to my knees in the abject Indian manner and volubly. He stopped his exercises, and came up to me in the Hindu gesture of total humility. "Please, sir, I have no chance here. I will do anything you ask. Please get me to America."

Samir watched me watch the yogi and I watched my mind. I felt helpless as a child to do anything for myself, even to end this scene where the dark glowing eyes energized the jumbled content of my mind, showing me more clearly than ever before, the hopelessness of my attempting to do anything on my own because there was no longer any my own, simply a conflicting montage of successive inducements for my body to act in contradictory ways.

Samir said, "Come." I followed him out the temple to the ghats that led down to the water. The yogi tried to follow us, clutching at my hand. I saw my body turn with a violent seizure. "No," it said, "No, No, No!"

The next day we arrived at the end of the road for vehicles, Rishikesh. Above Rishikesh pilgrimages continued on foot. The Indian Army had closed the road to vehicles other than those it drove forward to confront the Chinese. A row of five lepers held out their stubbed dwindling fingers. The City of the Rishis, of the immortally wise, and of the fatally ill. And all those like myself searching, scrambling between the gorgons and the muses, between the wrathful and the benevolent deities, asuras and devas, fear and attraction. Samir would not let me stop in Rishikesh. "There are no rishis here any longer, we must go onward to Badrinath. That is our program." This was the other half of Anil's teaching: every day he made up a program and then followed it, nothing but death or jail could stop him.

We walked, loose-jointed, up the road to Badrinath, dusty, fewer and fewer people, rice fields terraced up the mountain slopes. "Otherwise it's not dharma, only a habit, "Anil replied. "A dharma a day," I had joked him, "keeps the doctor away."

The Gunga ever more torrential, louder, and boulder-rolling to our left. On the second day an Indian army patrol turned us back. Rumors of fighting in the mountains with the Chinese. The afternoon sun low on the horizon. Samir set off, zigzagging up the mountain on the little dams behind which the potential energy actualized into rice paddies. Night. The Himalayas cut a vast swathe out of the stars. We continued following the star glimmer on the bare dam paths, hardly able to see our way, until at midnight, exhausted, we arrived at a very small traveler's shelter about two thousand feet above the road and about a mile and a half east of it.

Samir unwrapped his chapati and dal and we ate in silence.

Samir made no effort at personal chit-chat, or at philosophy, or at politics; we traveled, we rested, we ate in silence. Whatever he was doing was at Anil's request to do for me. Anil, I knew, was a karma yogin, a man who used action as the way to attain, and as the way itself. Samir kept up the action.

At sunrise we arose, washed off in the chill paddy water, and after eating more of our chapati, descended again to the road. "Since I cannot now take you to Badrinath, I shall leave you at this ashram. The master here is genuine."

The ashram, composed of one story whitewashed buildings, lay inconspicuously on the bank of the Gunga, above a wide field of rocks and boulders, deposited during high flood times, that led slopingly down to the water. Samir had to talk earnestly, supplicatingly, authoritatively in turn with the senior students of the master for about an hour before they agreed to let me stay for three days. I was to observe, to be at my freedom in the ashram, and to have a sleeping space on a twine cot in a room with two cots. For toilet, one went out to the rocks along the Gunga. "Every flood time, the Gunga cleans our latrine area," a student told me.

By now, my organism had adapted perfectly to the dal chapati diets and daily walking of miles. I had no extra weight, but was not gaunt, either. About one hundred and seventy pounds. I enjoyed perfect condition because though the athletics I engaged in were mental they moved whole systems of muscles in subtle and powerful articulations without any intention on my part. I visualized the plot lines and chief actors of *Mahabharata*,and the *Ramayana*. The afternoon disappeared into the whitewashed walls, the sunshine, the troubling and contrasting figures of Sita and Draupada, Ravana, the army of Hanuman, then the battle at Kurukshetra, and the vision of Krishna revealing the omnipresence of infinitely formed divinity demanding performance of duty no matter the cost. But no Krishna appeared to give me my orders, no Ravana appeared, an overt evil to fight. Perhaps I had already fled from my Kurukshetra, my Ravana, the battle for art and truth and freedom, the battle against conditioning and prejudice in America? But there had been no battle there, only aspirations. I could only ready myself for when the battle came. Perhaps the principle called Krishna would appear in some other form as my flight commander at the right time and right place with the right people.

Going back to my cot under the stars glittering with light, flavored by the sound of rushing water, I found a sturdy blond mesomorph sitting in full yogi position.

"Jean Paul," he said. French.

"Joe Madison," I said, sitting down in my relaxed crosslegged position. We looked at each other for a minute or so.

"I will tell you my story first. And then you will tell me yours."

I nodded, caught by a sudden bond to this confident blue-eyed yogin. He knew something I didn't.

"My family have been merchants of wine, of pears, and of apples in Les Halles in Paris for centuries. We are famous for our taste. A pear, for example, is at its peak for only one day, and we select and deliver the pears to the high restaurants of Paris. I never taste a wine, I decide to drink or not to drink by its bouquet alone which tells me everything I wish to know."

"I enlisted in the Algerian war and became a commando. I became better at killing than at tasting. At first I did it for the honor of France. One day I realized I killed because I liked it. Finally I grew to like it—too much."

"I came to India three years ago, 1961. I had secured the name of a man in Mysore, unknown in Europe, from a friend in Cairo. I thought for a long time how to introduce myself. At last I found a rug which possessed everything that I could see at that time in the way of form, color, texture, line, and symbol."

"I came to his house and presented him the rug. He contemplated the rug. After a few minutes he asked me to stay. I learned, gradually, how to pull out my entire lower intestine and clean it daily with a toothbrush."

"After that, when sexual desire threatened to undo my work, I learned to sit in a tub of water chilled to freezing with ice, and draw in sufficient water through my penis to discourage all organic interest in sex for the next twenty-four hour period."

"At last," I said, "I can do it, I can master my body and live for a hundred years and more, but do I wish to destroy my sexuality?" I left this master, perhaps the supreme physical adept living today, and came to the master here, a mental adept.

"When my mind stabilizes and I have made the Stone, I shall walk back to Europe, over the Khyber, then Afghanistan, Persia, and Turkey to test the power of my work, and after I shall confront the vast confusion of the modern day. If I can do that, I shall know the Stone exists and I will become immortal."

"You will visit me in Paris when you have done your work and then we shall see what gold-making fire can be ignited when two stones rub together."

Of course I had fallen in love with this handsome daredevil who spoke so simply, so classically. At the end of his discourse, my body had lengthened to total erection, my neck had straightened out as never before, the mulabandha had flexed my intestines back flat against my spine, the air seemed so dense my breath scarcely moved, nor hardly did it need to move the air was so delicious and nourishing.

I told him of the peyote trip. "I knew the pilot in that B-52 would drop the bomb upon getting the order. I knew the President could give him that order. I knew then death had been set upon us not just as before as separate organisms, but upon our species, our evolutionary post-species, time and space, cosmic potentialities. I made up my mind to search for the way out. Nothing mattered beside that."

I told him of Tangier, of the Nile, of Anil. The next day and a half passed by like a minute, like years. We parted only to sleep five or six hours. We walked by the river. We sat. We hardly talked again. We communicated being. Anil had been far above me, the faces from the crowd too low, too predictable, even their anomies and schizophrenias. Jean Paul and I had touched wings in flight and now I knew I was not mad. It was possible, though, that neither he nor I might make it. At the moment he seemed to me slightly ahead.

He went to Rishikesh late in the morning of my third day there. I could see the senior students turning against me, innerly, because I exhibited less interest in them or their master than with Jean Paul. I endeavored to emanate my gratitude to the master for creating this place of inner freedom where Jean Paul and I could meet. I realized he was master of the mind, the fifth yoga, only one below being master of the theater of action, Anil's status, but Jean Paul and I, like Samir, still struggled to master the fourth yoga, the action in the inaction, the inaction in the action. But I imprinted the taste of the balanced life of the ashram, and knew that following this taste would lead me to the source of mind. The mountains, the river, the boulders, the heavy sunshine, burnished the ashram's dharma.

In the middle of the afternoon, hullabaloo. Jean Paul ran, sweaty, past me, throwing his few things into a knapsack. The senior students crowded in back of him.

"In Rishikesh, those dogs. They bark and bark at you. Maybe they smelled my killings. Maybe they bark at everyone. One came so close, barking, I thought he would bite me. My reflexes. I left Mysore too soon. My reflexes. Suddenly he was dead. Only then my body stopped and I looked. A broken neck. Rishikesh, all life is sacred. Except when you kill an animal. Then your life is no longer sacred. They all came at me, almost at once. I almost started to kill. They would have had no chance. How I hated these fanatics. They would wreck all my work, even kill me because of that dog, dead through no malice, dead by my killer reflex, faster than thought or emotion."

"I saw the horror. I stopped myself. And then I ran for my life. I, Jean Paul, ran. I ran until I outran them all and most of all I outran myself. Who am I now? Who can say?"

"My life in Asia is finished. I return to Les Halles. I will see if I can take up the work again. Come see me. Help me. Goodbye, my friend."

He disappeared quickly out of the ashram and with the double skills of yogi and commando, fueled with adrenalin and remorse, almost disappeared before he disappeared over the ridge across the road. Myself and the senior students stood, outwardly impassive, watching long after we could see nothing but the ridge and the sky.

Suddenly drawn to the master, I sat down near him, where he was compounding perfumes and medicines from local herbs with four of the nearby villagers. Night fell, and the villagers withdrew, their faces radiant, carrying their small packets.

The master arranged his jars, and recrossed his legs, entering the motionless state.

"And this is all he does?" I asked one of his students. "Sit, walk a little, compound perfumes and medicines. You do all the work, it seems to me, to make the ashram run."

"Yes. You see, we live still such complex lives. It is true, he gives out no teachings, he does not give us many exercises. But he has simplified his life, and that is what we are studying to do. Around his simplicity we all revolve."

At the end of the third day the students insisted that I leave since I had found no task to assist the ashram. In truth my life while not so simple as the master's, was simpler than the students. Jean Paul had given me much, perhaps most of all the vision of how easy it could be to lose all one had made if one had not finished the work. Now I could not stop, I could not even continue as before. I had to accelerate.

If not to the Himalayas by Badrinath, then on to Katmandu which had opened only two years before to visits by Westerners except for the occasional individual by Imperial posting, or by special approval of the king.

Nepal

FIVE OF US lay comfortably on the stacks of steel rods in the back of the overloaded truck tooling its way up the just-finished Indian built highway that for the first time allowed direct access to Katmandu from the plains of India. The Sikh driver charged us twenty rupees and we climbed up on the fender and hauled ourselves in over the high wooden sideboards.

No one talked much. The excitement that led us on was completely inner. None of us had met before. We had all showed up at the nondescript border post as, from what we were told, from one to eight people did every day. No couples, no groups. Individual seekers.

The woman, slim, elegant, that lay next to me finally intrigued me with her relaxed concentration.

"Oh, I'm the secretary to the French Ambassador in Delhi. He has given me a month to make my studies of the Tibetan art. I have an introduction to a Lama."

We were in the mountains, actually only the foothill ranges leading to the Himalayas, when night fell. The truck pulled over by the side of the road at a two story stone house with low ceilings.

We drank tea and ate the simple rice and chapati meal, then climbed the wooden ladder steps to a room where we would all sleep. A faint moonlight penetrated the small window.

Marie fell to sleep quickly without a trace of coquetry between me and a nineteen year old Frenchman. A German and a Dane slept on the other side of Philippe. Dreamless sweet sleep. Early wake-up and tea in the false dawn, greenishly floating into the dim red bricked house through the open door.

Arriving in Katmandu, walking into a temple, bronzes, carvings, musicians, songs of God, hashish given me, lying back on the stone steps, clouds crossing the temple space interrupted by blue, all day, the night came, stars, the musicians stopped. I walked away from the first experience of civilization I had ever known, and realized for the first time that structures of being existed beyond individual enlightenment. The master's ashram on the Gunga, dependent upon his state and presence was only the last remnant of a great civilization in which one time his school would have balanced as I had balanced in his school. Now the religious mobs, pushed onward by political masters, washed around his outpost. I realized that going nowhere from this place led to fuller and better than going there from the Kief café. I knew now not only what I sought but where I wished to live. High civilization. The musicians smiled at me as I left.

An Indian student came up to me as I paused at the temple entrance wondering where to go. "You can sleep in my room. There are no hotels here." I slept curled by his small desk at the head of his cot.

Tibetan refugees filled the hill courtyards above Patan with their red and yellow hats and robes and the city squares filled with the annual heaps of yellow grain and strings of red peppers. Amid the ceremonial throngs, and the entranced musicians, I studied the carvings of tantric ecstasy, where sex itself transformed into the motionless state in which as pure energy it rose into the brain, memories of insights beyond the senses had enabled the artist to create equivalent insights that he could relaunch from objects limiting to the senses themselves: objective art, mathematics that from the chaos of the senses brought forth form to harmonize the life of home and street and field.

I knew the Tibetans at once. I had seen them and lived with them before. Cheyenne and Hopi. Bodies lived in like a cave out of which intelligence prowled from time to time on dangerous galactic hunts. The Tibetans were Indians whose Mexico City had only now been destroyed by the new savage Conquistadors. Suddenly the insight: I would find the next step to the secret from among them, however it might be hidden. Weaving their rugs, their legs straight before them on the ground, the warriors looked like all the great captured tribesmen of the past, proud, incandescent with memories of manhood, conscious of their tragedy. They had fought till the last moment of slaughter when modern weapons of army, state, ideology, and economics had forced their leaders to surrender or to flee their homeland.

At Bodhanath, I circled the great white stupa. Marie appeared with Philippe. "I am going to see the Chini Lama. Do you wish to come with us? He will perhaps do something with us." She accompanied the words with an engaged, exact smile, emotionally precise. Sex exploded between our foreheads, two feet apart, and erotic energy pulled the debris together in a secret determined enterprise. Phillipe saw nothing of this.

Chini Lama smilingly, gravely, took Marie's letter from the Ambassador. We sat in his second story room looking across to the stupa, one of the central pilgrimage points of Asia, and of which he was the master as well as of the adjacent gompa. His nephew brought out piece after piece of extraordinary art. Books of layered loose sheets, written in gold, written in silver, carvings, jewelry. Each refugee had tried to carry out of Tibet one or two or three things of value, across the passes, through the battles, and, penniless upon arrival, had given them to Chini Lama to sell hoping for a few rupees, and then gone on to the refugee camp in order to survive.

Marie and I both wished for the same object out of all those presented. Ten dollars American. A wooden three sided board that was a cookie mold, to mold magical cookies. She wanted it for a French museum, I knew a man in New York whom this object would galvanize, and who would place it where it could influence many people. The Chini Lama placed it aside with a laugh.

"You have each expressed desire to experience the real Tibetan teaching. I will send you with this fourteen year old boy to the Lamasery in the high ridges, about thirteen thousand feet. Make this journey and you will see whatever it is now your time to see. After that you can decide who will take this piece."

"How can you keep such extraordinary cheerfulness," I asked him. "You seem to me perhaps the only completely happy man I have ever met, yet your country has been overrun, the centers of your culture and teaching destroyed, your people impoverished, you are selling masterpieces for almost nothing. How are you able to do this?"

"Our teaching is that the universe returns at intervals to the diamond point. We have returned to the diamond point. We are not destroyed. From the diamond point, new and greater manifestations will occur."

I looked at him and he let me examine. I realized I had met my third teacher. Anil had shown me the way of action: being confrontation, and of the theater of action, a new program each day that must be followed no matter what difficulties arose. The master on the Gunga had shown me the way of contemplation: a simplified dharma based on providing high quality services to the surrounding community so as to be allowed the time-space for cosmic thought. The Chini Lama had revealed to me the start of the way of the dramatically dangerous universe itself: find the diamond point, then one can lose everything, and recreate the next world, immortal in the midst of catastrophe, beyond animal, president, man and god.

The conversation broke off when a man entered, and began an excited dialogue with Chini Lama. He left, backing out on his knees, after kissing Chini Lama's bare ivory-colored feet who had no more change of expression at this event, than at our arrival, or during showing of the objects for sale. His face remained open, affable, cheery, sparkling-eyed. Many people considered him something of a trickster.

Chini Lama accompanied us to the space behind his house. The boy placed his heavy open back leather pack on a raised ledge, then leaned against it to place its strap around his forehead. "Follow him," said Chini Lama.

We began walking up the road behind Bodhanath. It soon became steep. The boy easily kept his lead. A two-inch diameter pipeline carried water over the trail, splashed into a paddy network on the other side. I gulped the cold fresh clean water. The monsoon had not yet ended, it was September, and the water trickled and ran in myriad sparkles throughout the flashing greenery.

On the high ridges, the houses disappeared, and we began to wind through dripping trees and over small streams. Leeches appeared. Our fourteen year old leader would, without breaking stride, reach down, pull them off, and throw them to the side. He was barefoot.

The leeches wriggled in through the eyes of our boots, fell onto our arms, crawled down the hat onto the neck. At stops we pulled them off, gorged with our blood. Our socks began to turn red.

Great waterfalls thundered in the distance. At night a mother and her eighteen year old daughter put us in their stone house, the men away working in town. The yaks shared our quarters, separated by a low wooden fence. I slept between Marie and the daughter, all five of us lying on the green floor of the mountain house. We had eaten delicious yogurt made in the long three inch diameter wooden yogurt pipes, held together by silver.

The idea crossed my mind that I could sleep with either Marie or the girl. I watched the sexual struggle and all its endless arguments, desires, fears, take place. Half the night I lay awake, but without becoming or rejecting sexuality, using the energy of the delightful alternating thought forms of the sophisticatedly beautiful European and the intense glow of the young Mongolian. I realized then that sexuality was energy and that it could take me nowhere instead of any old where. I let my thoughts expand to include the young Philippe, the round breasted mother, then I realized the yaks were also awake now staring at me over the one rail that separated us from them. I gazed back into their dark eyes lit by a faint moonlight. Then earth, yaks, people, myself all united in a living swell that crested into dreamless sleep.

Early the next morning, before dawn, we started. The Himalayas towered above our thirteen thousand foot high trail. Clouds arose from the valleys and raced toward the sunlight a thousand feet a minute until they would cover the great peaks themselves. The waterfalls boomed more magnificently.

The third day we arrived at the Lamasery. I could see at a glance that the man who had charge of the temple was not the man who had built it, nor even the man who could understand it. I sat myself down to study the yantra, the symbolic paintings that under proper concentration and visualization could become an action, and to experience the space. Philippe and Marie left to study the small village nearby.

My two great Tibetan heroes looked at me. Padma Sambhava and Milarepa. Both masters of magic who became masters of themselves, the two creators of Tibetan Civilization. The man of passion and the man of thought. Why did Milarepa not complete the tests set him by Marpa? Why did he accept Marpa's wife's help to make it easier? Was that the flaw through which the Chinese had entered, destroying it back to the diamond point of Nagarjuna, the absolute nihilist? Above the Yab-Yum of equilibrated sexuality, the two images stared across at the wheel of the reality of life, and between these visions the apprentice Lamas and now myself sat, eager to attain to the evanescent lightning energy filling the sky of the mind that could turn into the small dazzling diamond at the gate located between the cerebrum and the cerebellum. At nightfall Marie returned. I had not penetrated past symbol but the symbol was beginning to move. She began to move around the painted space, studying the yantras in intellectual detail, perhaps readying for a discourse at a museum. I could feel my fledgling being collapsing, her display of brilliant scholarly intellect sounded like a droning machine. I felt the desire of sex rising like revenge to dominate consciousness. I began to fight against her image. To my horror, a deep hostility as when great danger threatens, invaded me. It worsened. I became excessively polite to her, attentive, wishing she would shut up.

The Lama and his wife, red hat Nyingmas, gave us a hospitable dinner. We slept. I made myself not touch Marie. I could not stop the hostility and the wish to use sex as a weapon.

Returning, the rain began when we reached a collection of four houses about halfway back. The boy had remained at the Lamasery and we had no guide.

"We must move on quickly and reach lower ground and the main trail before the big downpour starts," Marie said. "We will get lost here or suffer extreme exposure."

I felt a terrifying force enter my body. I knew, no matter what the cost to myself, I must oppose her. "I think we should stay here for the hour stop we planned to rest," I said lazily. I knew she was right. I knew the rain would be monsoonal, intense. I knew it would be dangerous to stay. But I wanted to get rid of her. I wanted to love her, and now not totally in revenge, desire flamed in me also. And I loved Philippe, but something in me, not I, something greater wanted — an ordeal. The hostility in me must be destroyed and I could see no other way. I could see myself acting stupidly from any rational point of view. I knew we should stick together. The fearful part of me pled with her, "Marie, let's stay for only half an hour. Then let's go."

"No, it will be too late. I am going now. Philippe?" Philippe looked at me. He would have liked to stay, but he didn't dare. I wanted to shut off them both. She turned her gaze from me abruptly and strode off. Philippe followed, giving me a last wave before disappearing from sight. I hated her for leaving me alone. I hated him for deserting me for safety and a woman. I exulted that I was alone. Now I had to do it by myself. I would not beg a sleeping space from the house people.

I had four small potatoes left in my pocket. After an hour I got up from under the overhang from one of the small buildings. A woman and child watched me leave. I took a last look at them, framed in the doorway.

The sky was darker, the wind up a little, the rain chillier.

Something had taken me over. I obeyed it. I wished to learn something. I could not, would not go on the same way. I remembered Chini Lama saying "You will see what it is your time to see."

I began to see many forks in the trail. Coming there, following my guide, I had seen only a trail. Now small trails were frequent. In my present state, they all suddenly seemed equally significant. On each trail would be found a destiny. Who could say which was more important? But I knew only one of them would certainly save my life.

I began to wander. Suddenly I knew I was wandering. I could see the general line of the ridge, I would avoid the deadly sides where slips and avalanches occurred, but I was wandering back and forth.

The heart-jumping growl-bark of the deadly Tibetan dogs, trained to fight even tigers, brought me to my knees. I knew I should make myself remain motionless, but this motionlessness was inflicted on me, no strength remained in my legs. The adrenalin had gushed out in the rain.

The dogs' savagery showed me the utter jackassness of all my unguided actions. Voices came up, two unsmiling men led the dogs away, and waved me on in the Katmandu direction.

The wind became lashing cold, the rain torrential, I could not see more than ten feet ahead, night almost fallen. An old tree with a cavernous trunk. I huddled inside, body aching, chilled, methodically raising and lowering my feet, working my fingers in my thin khakis which I had taken off and wrung to dampness and then put back on. I ate the four one inch diameter potatoes.

Towards the end of the night the rain stopped, but I shivered till the light came. In the light I moved in the direction of Katmandu, as near as I could make from the shape of the mountains, watching again the clouds bolt up from the canyons to cover the peaks. The day went on and on, I became weak with fatigue and hunger. I had missed the trail through the paddies. Perhaps I had even missed Katmandu valley. But by night I was descending. A river valley. Boulders. A sandy strip. Worn out, I pitched to the ground. Lying there, I wondered vaguely if I could ever arise. In the morning I awoke, almost dehydrated, the water not far away but I couldn't move. I had hit myself on falling down. The morning wore on. I was slowly concentrating all my strength toward an effort to rise when—laughter, voices.

Some Newari from Bhaktipur had come up the valley. They turned me over. Tea. Chapatis. Life. Had it been the diamond point?

Marie and I sat in front of Chini Lama. "Alright ", she said definitively to Chini Lama. "I agree he should get it." We had both told our stories.

She stayed to buy several items for France, cool, beautiful, a woman such as I had never met before, whom only a year before I would have done anything to attain.

However, I had but one aim left me, to complete my voyage around the Planet Earth and to ascertain for myself its true shape now that I had created a point in nowhere from which to see.

Viet Nam

FROM NEPAL I'd decided to return to the world, not search again or deeper for awhile, to digest, to discover what the world would show me now that I had found a new place to be, a new eye to see.

In Calcutta I slept in the basement of the Gurdwara, filled with the white robed and bearded, woke to the four thirty a.m. chanting, walked the avenues of departed Empire and took a fast reprise of the ideas bedeviling the human mind in mercurial discussions with dialectical Bengalis vibrant in eighteenth century English expounding nineteenth century Marx and twentieth century film among the cardamom flavored teas in rapid service teahouses.

In Burma the golden spires of Shwedegon, man-made mountain on the Irrawaddy, the vast simplicity of religious commerce strewn along the great steps moving up each of its four sides, shop after shop existing from the stimulation of history, architecture, religion and cunning artistry; Buddha, craftsmen, and shopkeepers all profiting from the profit that kept Shwedegon crowded, maintained, and orderly.

Sickness, almost death, in a little wooden house in Bangkok, after being seduced by a lithe and handsome youth. I went through the motions with him, all the endless moral-psychological introspections of the guilt culture born into had disappeared without a trace. I see no particular interest in male sexuality in my organism, a neutral affect, a haunting question answered. Headaches. Fever. I make myself walk to the river and back. Take in sensation for a little energy, turn yantras for a little more. Can my new state now cure my body? Fantasies have begun to invade that state. After the fifth day of not being able to eat, I realize death is on its way. I barely haul myself to a nearby hospital. My fever is 104 degrees. The antibiotic cures me in two days. Definite limits exist to the powers in the new tempo of the organism.

A boat to Songkla hauling twenty young girls from the North to the newly opened bar, the calm waters of the gulf. They are friendly, twittering, but I sleep alone on the stern, warm body brimming with sensation I refuse to spill out.

Singapore. A Chinese apartment vacated because of death, given to me for a month to exorcise the ghost, a street composed of Buddhists, the Tong, and the Party. Everything was business and I would watch every day an hour or so the old ones about to die on their pain relieving opium, watching and listening to the hammers of the craftsmen on the coffins being made outside the door. The man living in my apartment had died too quickly for his family to send him there.

Gliding up the Mekong in third class on the Messagerie Maritime passenger vessel, I knew as soon as we pulled alongside the Saigon dock that here I would find the limits of what I had learned and where I had to go.

The ship had no sooner stopped, than two girls, each with parasol, bouffant, poked their heads through the portholes and quickly found welcoming arms that pulled them through and set them upright on their high heels. Within a minute of animated bargaining they disappeared into the neighboring rooms.

A German engaged in his wandering years, Heinz, and I strike off on our own from the ship. He had heard about the beauty of Vietnamese women and, of course, had procured an address from a previous German.

"Don't worry about your money," he said, "I know you have just a little to accomplish your purposes, but tonight I want you to go with me." He wanted a companion for safety, that was clear, but my policy, since Nepal, had been to take whatever came along to observe the changes that might take place in my life. It just now occurred to me that the adoption of the ability to carry out this policy was itself perhaps the greatest change. Before, I had always fought to stay intentional, to stay in control, to make it happen the way I desired. My life had disappeared in prodigies of work and effort that had never accomplished anything beyond transient amazements.

To my surprise, Heinz took us to a middle-class district of Saigon, composed of very well-crafted houses with steps going up to the entrance several feet above the ground. A girl looked at the letter and showed us through the living room into a master bedroom in which two Vietnamese men and another girl sat on a large bed, laughing.

After a couple of minutes, the two Vietnamese men left. One girl laughed and kissed Heinz, another myself. I thought they were both beautiful and charming.

Then they bounced off the bed, catlike and the older one began to bargain. Their laughing stopped as Heinz showed strong resistance to the price. Their voices became lower, pleading. Their eyes turned on a passionate look. They each got close to us, knees pressed against the bed.

"No. No." Heinz named a price slightly higher than his original. But he would not go up. The girls revved up the passion, the eagerness, the pleading.

Suddenly they quit talking and turned off their eyes. Heinz glared at them. I glanced back and forth at Heinz and them. He jumped up, cursing, and we strode out. On the stair, a fine big blue vase stood. Heinz picked it up to hurl into the street to break it to pieces.

I grabbed his arm with all my strength. Totally surprised, I watched myself hold this enraged man from breaking this vase. "I won't let you do this, Heinz." I could not believe my own voice. I had never stopped anyone physically before in my life except on three occasions when they were attacking me.

Heinz put down the vase. "But they deserve to have it broken." he snarled.

"But it's too beautiful to break."

His walking became feverish and soon we arrived back in the center of Saigon, just off TuDo street, at midnight, way past curfew hour. "We have to go back to the ship, Heinz."

Suddenly he saw a girl in Ao Zai slipping down the street. He ran over to her. She nodded and he waved me over to the front of a chic store that indented concavely about six feet from the sidewalk. A policeman in a white uniform caught sight of us and ran up to the store.

The German gave the policeman a haughty look and handed the girl some bills. The policeman and I watched him methodically fuck the girl, one, two, three, four, five, never changing his speed. When he finished, he motioned to me. I started to say no, but the girl stared straight into my eyes till she caught hold of my attention. The energy built.

The policeman put up his hand signaling wait, and dashed off around the corner. He came back in less than a minute with a big piece of cardboard which he put under the girl who braced her right foot against the window pane and invited me on. After a tour de force, Heinz and the policeman clapped, and I told the girl to meet me at the same place at noon the next day.

Heinz went back to the ship, but I knew I would not leave Vietnam for a long time. I loved it. Wandering down the street, hiding from the occasional patrolling jeep, I arrived before a Buddhist Temple.

I lay down on the step below the top, the night warm and windless, the concrete soft as grass to my organism. It must have been two in the morning.

Someone was lifting my head. I felt completely, classically, Asian and accepted whatever would come without a qualm. A pillow slipped under my head. A little later a sheet fell lightly on my body.

At dawn the worshipers began to arrive and I rose to join them in front of the Buddha. Upon leaving, an older boy took me by the hand and led me through a door, down a corridor and into the Abbot's office where he sat behind a polished hardwood desk.

"Many Americans think our Temple harbors hostility to them because one of our bonzes set himself afire to protest a policy. But we are Buddhists, hostile to no one, even if we oppose a certain policy."

"I have studied Tantra with the Tibetans and Hinayana in Thailand. I wish to study Mahayana with you." We spoke in French.

"I do not speak English and wish to correspond with colleagues in India. You can stay here at the Temple if you will translate for me this correspondence. And you are not studying Buddhism in the right sequence." He smiled.

Leaving the Temple about ten that morning after translating a letter addressed to Bodh Gaya, I bought a papaya, red-gold, sweet, delicious.

To my left an extraordinary geometric array of zig-zag trenches dug throughout the park. "What do these symbolize? I have not seen them in my study books of the teachings." I asked a man who looked as if he spoke French.

"They symbolize war," he said sarcastically. They are for us to jump in when the mortars and bombs begin to fall." The fact that a fool of the first order still lived inside me gave a greater shock than the resistless approach of death in Bangkok.

The girl met me at noon. She took me to her room, the first floor of a two floor small wooden house. She weighed perhaps eighty pounds and I thought her exquisite.

She insisted she did the work only to support her invalid father and crippled mother. What else could she say. Perhaps it was true.

That night I wandered into a bar near the river. Full of American G.I.'s. The Americans, I had read, only had Special Forces units in the mountains, some river patrol boats, some Helicopter units, some Air Force units, some Navy ships in the Gulf of Tonkin, and some soldiers guarding the Embassy and Air Force Units. There were agriculture advisors from A.I.D. and of course the spooks. So it was not too surprising to find a bar full of G.I.'s.

Well aware that my beard, khakis, and rubber thong sandals made me stand out I stopped short of going into the interior and sat down on a stool at the side of the bar about six feet from the door. The bar lady slid me a beer.

A girl of five or six emerged from the interior of the bar, selling newspapers. I bought one. She stood with her back against the bar, her head under the overhang, gazing at me intensely.

We began staring into each other's eyes. The soldier catty-cornered from me on the bar stared into his whiskey. Everything seemed to stop. My hand went down to touch her shoulder, but her legs opened. Our eyes locked. She rode up and down, squirmed vigorously and kept my eyes locked as my hand withdrew after the release of tension. She broke eye-contact and walked out the swinging half-door.

Sounds began to enter my ears from the back of the bar. "You better get out of here," the soldier said, "a bunch of us white soldiers came here to fight a bunch of black soldiers and it'd be a shame for your head to be busted for what after all's none of your business."

I swung around on the stool and moved out and down the street headed for a place I'd heard served great café au lait, The Hotel

Majestic.

The café au lait, the terrace, the boulevard hallucinated a tropical Paris and these right bank trappings in Saigon seemed as bohemian as the left bank in Paris. Stunned to the edge of my mind by the events of the last forty-eight hours Nagarjung's doctrine of absolute nihilism made mantra to my attention, "The world of phenomena is unproduced." "Do you mind if I join you?" Trimly bearded, in a light green suit wearing low black shoes with shoe-strings, Leonard McCallister dropped his skinny dynamic frame into the wicker chair next to mine.

"And what are you here for?" he asked.

"Here for?"

"Yes, nobody comes here without being here for something."

"I arrived two days ago on passage in a freighter headed for Japan where I intended to see Kyoto and study Mahayana, but I fell in love with this place. The ship's sailing out now, but I decided to stay. I have no idea what's happening here but I can feel that this is the high energy place. I want to experience it all. And you?"

"This is my second visit here. You are not the only one to be uncertain as to what is occurring. If I'm not, it's only because I also visit Moscow, Peking, and Hanoi regularly, and there imbibe from the fountains of certainty. I'm Professor of History at London University."

"How can you travel in all those places so freely?" I asked, feeling the envy twinge for the first time since meeting Anil. Not serious, like a fly buzzing, but quite noticeable. "I write a couple of articles a year for London newspapers explaining the Party's view as to what is happening, of course, not stating that it's the Party's view, and the comrades think it's worth my plane fares and hotel bills."

"You sell your mind."

"No. I always identify the remarks as coming from an official connected with the Party. Although I must admit about eighty percent of what they say seems about right to me. But it's an easy and amusing way to travel to places that otherwise a middle professor could never afford." He laughed frankly, after a few seconds pause. "It's a cheap price to pay."

I immediately liked Leonard. Brash, cocksure, driven by hatred of Oxford and hierarchy, he patrolled the edges of the Revolution, and he, at least, believed that Saigon was the way to London, if not to Washington, even if Beijing had failed to be the way, as yet, to Paris.

"The Achilles' heel of imperialism. The Americans were stupid enough to deal with DeGaulle instead of Ho Chi Minh. Of course they were between a rock and a hard place. The French would have turned against them. They've inherited the French fiasco and all the Annamite hatred of the French. They don't realize the Annamites, untraveled as they are, think the American Democracy is just like the French Empire. The Communist leaders have never lived in America and they believe it's just another European small ruling class structure. The Chinese and Russians will pump supplies so long as Vietnamese will fight, and the Vietnamese will fight for a century if they have to, they've already fought for over twenty years against Japanese, French, and Americans with no stopping. This is a huge country, a thousand miles from North to South, with the population of France, bordering on China so the Americans can never stop their weaponry resupply even if they blockaded Hanoi, and the Revolution is under the total control of the Party, which is under the control of two geniuses, one political and one military."

"Why won't South Vietnam be like South Korea?" I asked. "We won there."

Leonard laughed. "Do ring me when you come to London. I like you. Because, dear boy, the Party controls the countryside here, because this time nobody will back the Americans, not even the French, especially the French because they're the one's who lost it, and they don't want to be replaced, because your President understands nothing besides Texas, most of all because the Revolution has become more powerful."

"In the evening at the Majestic over my third café au lait and after my first International Herald Tribune since Bombay, I talked to the correspondent for the Far Eastern Economic Review.

"You don't want to go out to the provinces," I said, "it's too dangerous. But you need to get all the facts to write about."

The idea of war had seized my mind. How could anyone understand history without direct experience of war? In the Korean War I had built airstrips on the opposite side of the world, the Caribbean. I had to reach the front. Perhaps with my new organic state and hatha-raja-tantra techniques, I could see something no one else had seen before. It would be new for me at any rate. Even, perhaps, a necessary test of whether what I had achieved possessed any objective validity. I could already see my mind reeling at dangerous thoughts, thoughts that could at any moment have gone too far if coupled with movement.

"You could make me your assistant correspondent and I would file reports for you to select passages from."

A slim Englishman with blue eyes perched encroachingly up on his forehead across the top of which crept a deftly careless tuft of greying hair, a chin bent askew with ardent skepticism, he looked at me appraisingly. "Look here, I can't do that. The rules don't allow it. But, I will give you a letter stating you are doing some writing I'm looking at, and sign it with my official correspondent identification. If you're clever, you might do something with it."

The US Army guy in charge of correspondents, sleek and Venusian, looked at me closely.

"Well, this letter's not entirely regular, but you sound serious, and we'd like for somebody to get the real story from the field. But you've got to at least trim your beard and get rid of those sandals. Those boys in the helicopter units are way out there and they're very proud of their canteens. They try to keep the standards up. It means a lot."

"I'll get tennis shoes this afternoon and trim up the beard, down to an inch." I had always kept my hair from growing long anyway, as a protection against lice.

The green Caribou began tightly circling down over the Soc Trang airstrip. "We want to stay tight enough in so no fifty-caliber machine gun can sneak close up to the perimeter and get us." The circling was tight enough, alright. I could feel my intestines pulled sharply to the right by the centrifugal force.

Suddenly the barbed wire, sandbags, the choppers, the greeting, hustled to my room, warm welcome given to anyone who got out there.

Sitting on a bar stool at the canteen that evening, the guys on my left and on my right zeroed in. "So how come you're here? Don't you know it's dangerous? We're in the Service. We're ordered here. But you don't have to be here. You're crazy."

"Well, if you're covering a war, you've got to get out to the front some time. And I heard you guys are the best." "Yeah, so far we haven't lost anyone." They both rapped the bar sharply with their knuckles. "Some of the other units take chances. Let a chopper roll out on its own. They've got losses. We always buddy up. Somebody starts shooting, the follow guy takes him out. We work at it."

A colonel strode by, glanced, advanced to the lieutenant, and poked his finger against the crew-cropped skull. "Mister, your hair's an eighth of an inch too long. You will clip it tonight." "Yes, sir!"

The captain said, "I've got an interesting run tomorrow. Flying in medical supplies to a Chinese village. Victor Charley holds all the roads in that area from sundown to sunup. The Chinese are on an island and fight everybody off including in the past the Government and Cao Dai."

The Chinese leader advanced on the chopper as soon as the blades stopped. Gave a warm welcome to the Captain, shook my hand, beckoned us to tea, while his men moved silently forward as a group to pick up the medical supplies. All around the island, an expanse of endless green. Silence. We had seen nothing for miles around coming in. The village was a perfect target.

Our teahouse cantilevered over a canal that divided the island. "The Communists," he laughed, "have tried to send a boat right down this canal to blow up my teahouse from underneath but our machine guns got it first." His eyes glittered but his hands remained as relaxed as a trained mandarin's, which in fact he was.

"But you're basically surrounded. How do you keep going?"

"They know we will fight to the death. And that we know how to fight. They run a small attack now and then to test us."

About five hundred people originally lived in the island village; their number had swollen to a thousand from refugees coming in from villages attacked once too often.

We touched down at a Cambodian village on the way back to Soc Trang. The chief monks advanced in a line of three to greet us. Their acolytes pressed from behind for the medicine. Their village was deeply Buddhist, unarmed, but not because of religion.

"Neither the government nor the Viet Cong want us to have weapons. On this issue both of the Annamites, North and South, agree; the Cambodians must not be able to fight. You see, all of this area west of the Mekong was Cambodia until a century ago when the Annamites annexed it. The Annamites will not stop short of taking all Cambodia. The government, of course, is stupid in this matter. The Viet Cong have stripped all our temples of gold, they tax us. We would resist if we could. But the rich in the government think we might later on resist their taxation or their further colonization of the area. There are many like us who would fight the Viet Cong but the government will not arm us."

Back at Soc Trang, the doctor took me out to the practice range. "You never know," he said. "If you're going to be here, you should be ready to fight if there's an attack. Besides, it's fun.

He had brought out four hundred rounds of ammunition and six grenades. We had two grenade launchers and two M-1's. In addition he had two forty-fives. He brought cotton wads for our ears.

We fired away with great good will.

"You've done this before."

"Yeah. Korean War, basic and advanced basic, sharpshooter medal."

"But not the front."

"No, not the front."

"Hey, here's some tetracycline. Give yourself a prophylactic right after. You can't be too sure."

"No, you can't be too sure."

The town was off limits to the US Army, but I strolled down its broad main street, peaceful in the heavy brilliant sunshine driving the green chemistry that built the boulevard's dense glamorous visage rouged with red and purple Bougainvillea.

Invited into a cool shaded hardwood floored house by a local doctor. Tea appeared on our table while we sat and talked in French.

"You see, we do consider ourselves the superior race. For excellent cause. As an example, I was raised to be a Confucian scholar, but was taught Buddhist theory and practice of mind. Then I studied Western nationalism and scientific thought and became a doctor. For awhile I became a Marxist to understand the mass movements breaking up our old world. Now the American ideas of market and dynamism interest me."

"In the States, people call that going through phases. You know, 'now you are in the Dostoyevski phase,' it's somewhat derogatory both to you and the subject matter."

"Exactly, that is how you Americans avoid comprehensive thought. But you see, I find no contradiction in all these systems of thought and behavior. We do not go through phases. We integrate vectors and levels. Indeed, I believe more, perhaps many more exist than I can yet cognize. Nor do I hold as your liberal thinkers do, that these skills contain equal validity at all times and all places. No, to me each represents a facet of the truth, and each must shine at its appropriate moment upon the key aspect of the situation. The whole aim of the superior life is to acquire these basic tools and to use each at the right time and at the right place with the right people to produce the right action."

The rosewood chest sitting by the opposite wall was boldly, exquisitely carved. "My great grandfather's." French patisserie had been served with the tea. As I left, he shook my hand. "My friend, you can walk with perfect freedom here between eight in the morning and eight in the evening. After or before that and almost certainly a grenade would terminate your present existence."

We smiled at the door, and I walked back towards the base. Carpenters, bricklayers, and concrete pourers engaged in construction on one or two sites in every block. A boom. Textbook capitalism.

And at night the Viet Cong collected taxes on the new prosperity.

The next night, attracted beyond the possibility of resisting, I found myself lurking in town as evening fell. I had told the doctor at the base not to expect me back that night. "Covering a story in town."

A pedicab bicycled up. "You get in, quick, please. VC come quick. You probably dead. Number ten. You come with me, quick, please. Number one."

"I have no money."

"You no worry. Come with me."

We bicycled down suddenly deserted streets. The boom had stopped for the night. It, too, was sunlight driven. Dim lights from the houses.

Up the back stairs of a two story building.

"Madame good friend of mine. She give you girl to stay with

tonight. No worry. No money. You my friend. I like you."

Madame gave me a good looking over and then made a low call. A girl, about twenty, appeared smiling. Madame and my pedicab friend walked toward her room. The girl and I climbed in a small bed with mosquito netting for walls and met in a timeless place beyond class, nationality, and words.

The light of the sudden dawn appeared. "We go quick now," he said. And he dropped me off at exactly the point he'd picked me up, pedaling off with a cheery smile. No names, no addresses, no promises. Standing there, suddenly hotly caressed by the tropical sun, knowing that I existed in the middle of an unknown country, knowing that I had no idea of what would/could happen next, knowing that I didn't/couldn't know and would only continue to exist by some unseen alliance between my wits and the forces, benign, malign, and indifferent, that lurched and lightninged throughout this time-place, suddenly it all separated. The phenomena reflecting the garbled synthesis of my senses and the world moved by me like a far-off stream of water while I stood in a crystal solid that held my body immobile without awareness of effort. I experienced the intense force of the rush of the stream of water trying to pull me back in and then sweep me on its way. An intensity filled me and I let it rise until it filled me full. I let it work by being there and exerting its density. I felt it healing the me who stood there and I knew this healer was myself. "Me, myself, and I" started off the old secret oaths my cousin Bobby Lee and I used to take before the clod and cornstalk fights on my grandfather's wood strip at the edge of the grain field. I had liked "Me, myself, and I" because it seemed such a solemn frontier vow, but now I saw it had been a truth lugged in with the axe, the rifle, and the horse. Under that hot heavy brilliant sun which had disappeared from sensation and memory I stood in my crystal, holding my body and let the dense intensity heal me. I had been sick, and now I watched my healing. I could at last admit that I had been sick because now I knew how to heal myself. Oh, God, I had been so sick, sick to my very heart, and I was still sick. I stood there, crystalline, and allowed the healing.

"I'm going back to Saigon," I told the captain the next morning. "I've got my story."

"You're in luck. The whole outfit's flying up for a major strike against Victor Charley. You can fly up with me tomorrow. You'll see how we really do it."

Green lighted grasshoppers, we took off in a thrill of noise and coordination. Now I saw what locusts and geese experienced as the flock ascended, paused, took off, evenly spaced, headed irresistibly toward a place. The winged flock, higher metabolism, quicker sheer than the hoofed herd or the finny school.

The machine guns poked down toward the trees rimming the emerald padi. "How can they escape from this?"

"They freeze if they hear us. Don't move. Throw themselves down, like a footpath in the rice field. The Air Force call us Low and Slow, but if we get too Low, too Slow where we could really see, they'd shoot us down. At tree-line we're making 120 miles an hour."

The next day a hundred choppers on the pad. Eight Arven government troops poured into each chopper. "You get up with the colonels. Red Klaxon of the Chicago Trib with you. He's a real one. The only one of those s.o.b.'s but you that ever gets out to the front."

Red was legendary. I'd read his dispatches. He in his close cropped red hair, khakis, with journal and pen, jaundiced and observant eye. A veteran's walk, like an actor's, anonymous but ready to flash into individualized action, in his case, writing.

"Mr. Klaxon." "Hi." He never bothered to ask my name but gave up the energy to make one shrewd twinkle. I got it. "I know why I'm here but why'n the hell are you?" he said.

In India they called it prarabdha karma. Even if, a very big if, a man reached a stage of enlightenment as to the nature of reality, the momentum of his previous illusion-delusion mix would carry him on its resultant behavior patterns until its energies were exhausted, even if, an even bigger if, he stopped renewing its fuel. Homer, Tolstoy, Kipling, Hemingway, the movies, curiosity (that killed a cat), glamour, hatred of fear, who knows what else pile of not necessarily at all related vectors from past momentums drove me here.

We strapped in. The colonels, full and bird, made their introductions. They liked having two correspondents sitting back of them. The takeoff breathtaking as all the choppers rose straight up in unison.

"Fuck it!" The colonel yelled, "the radio don't work!"

"Fuck it!" The bird colonel sweated, "the radio won't work." He worked the instrument in every checkout mode he'd been trained in.

"Well, we're not going to call it off just because we're out of communication are we?"

"No, there's twelve hundred men ready to go and those eight hundred Arven have got to be landed. They're all briefed. It's orders."

With a fierce satisfaction the colonel revved off. Red flipped a glance at me. I got it. His eyes said, "this is the way it always is, you didn't know that did you? I knew you were green but you're gonna learn quick."

Magically to me, the fact that no communication existed between the commanders and the other choppers made no noticeable difference. We flew out in the tropic quick transition to dawn, a one hundred mile per hour technical city of green lights bearing the raiding civilization of genes and memes worked out millennia before the city. This was the Cavalry, the nomad tribe, out to devastate the sown. They knew all the moves both in the up to dated brain and the age old instincts.

Suddenly gunships were racing below us along the tree-lines. Machine gun and rockets. Behind suddenly great clouds of white smoke following the gunships. We reversed course and covered the ground again.

"OK, we're going down," the colonel shouted.

"Why?" Red asked.

"No communication," the colonel grinned.

Choppers everywhere landing, the Arven piling out into the padi, falling flat and then running toward the trees. The smoke covered nearly all the view. The colonel brought the chopper down to a landing in the padi, exactly the same looking rice as that back of the temple on the road to Badrinath. The tension rose in the chopper as we slowed up and landed. The trees looked very close and dark through the smoke. Red looked furiously and glinty in all directions.

In the chopper across from us we could see the Arven refusing to leave the chopper. They were clinging to the door. The Americans were kicking and pushing them. One by one the Arven dropped to the ground.

"Eager to get out, ain't they?" Red said abstractedly. The machine gun fire finally forced it's way into my visual trance. The world of light was thick as a solid. The Arven stood, squatted, and lay in the water and green padi and light like jewels scattered throughout a ball of translucent wax. Suddenly my ears nearly crumpled with the machine guns and rockets and phosphorous grenades exploding. In the midst of this extraordinarily solid world I saw how fragile my body and my chopper were.

"Let's get out of here," said the colonel. We roared off toward Saigon. The eight hundred Arven had disappeared into the tree-lines and smoke. "What happens to them?" asked Red.

"We got'em here. They'll clear the area and get back to Saigon on their own."

Red flipped another glance at me. I got it. No telling what would happen to them and the colonels would never ask. We walked away from the chopper on the concrete runway just like we had come back from an ordinary day's work. Everything seemed in perfect working order. No choppers had been lost. I walked all the way to TuDo stopping every click or so to buy another papaya to eat. Astounding, how delicious those cheap papayas.

Before leaving Vietnam I felt a necessity to visit the Special Forces base north of Dalat, near the Ho Chi Minh trail by the Cambodian border, behind which lived, non-existent only to the American press, and therefore to the American public, the North Vietnamese troops, entrenched, and supplying the Victor Charley units with weapons, instructions, and refuge. They realized perfectly that the Americans were victims of the phenomenological world, that is, appearances, and so would stop their planes and even the Arven soldiers right in the middle of the forests where some French cartographer had drawn an imaginary line separating nations. Of course no Cambodian army, customs, or anything else of that nature existed within twenty-thirty miles of the line. That area belonged to the State of North Vietnam whose army and commissioners occupied it along with some tribal people on their defacto reservations. The Americans would simply stop themselves and the Arven at this line over and over again and so no Victor Charley setback could ever end in a defeat. Whereas the North Vietnamese armed by the Soviets had to advance over the line to the sea only once for total victory to be theirs. No matter how brilliant the tactics of the Americans and Arven, their strategy doomed South Vietnam to eventual destruction when the Americans in the course of time would withdraw from the hopeless back-and-forth.

While the situation continued, however, all of those tribal people who could moved to live near American Special Forces Bases for protection from Victor Charley and to the degree possible from the South Vietnamese land sharks. Seven refugee villages of Ko Ho clustered within two miles of Sergeant Jackson's campground near the Dampao, the river of dreams, the last of the people with the small crossbows and bronze gongs.

Eight special forces sergeants held the base. Each had three military occupational specialties; for example: Sergeant Jackson specialized in Heavy Weapons, Communications, and Medic. Between the eight of them, every specialty of modern warfare was represented.

"You see," Sergeant Jackson explained to me at dinner the day I arrived with him at the isolated special forces post in the supply jeep from Dalat, "this is like those European wars of chivalry of the Middle Ages. No big armies here. Us against Victor Charley. Ee-leet against the ee-leet. They take a break back across the line and I go to Okinawa where I set my little girl up in a nice house. I told her she can have her lover or lovers, only when I'm there, I'm the only one. She keeps it just the way I like it, clean, simple, good food and beer."

"How long have you been in war?"

"Twenty-two years. Eight more to go to retirement. Nineteen Forty-two I hit the South Pacific, wound up on five different islands, including Okinawa, then Japan, then China, then Korea, then European maneuvers, then here."

Jackson's red hair set off his cool blue eyes, freckles, and nose that appeared to sniff the air. A curious deep faint sound arrived to my ears. Whump.

"Incoming mail!" Jackson shouted and we all poured down the tunnel entrance at the side of the table. Down below we waited a few seconds. Silence.

"Must've been a dud," Jackson said.

"Sometimes they just lob one in to keep us on our toes."

The next morning at breakfast Jackson told me, "We're going

on patrol. Now we don't take any unfit man. Hal's gonna take you on a little jaunt over those hills back there. If he say's you're OK, you go, otherwise you stay. We don't wanna have to carry you out."

Hal set off at a brisk pace. They had given me boots and in fact the whole outfit. Hal kept going at a brisk pace. We crawled down a sixty-foot tree that had fallen across a roaring stream. We pounded on through mud. We pulled and hauled our way up a steep rocky humid hillside. Hal was twenty-two. Interestingly, I could keep up. An objective physical test of the new physiological level accompanying the changes. I could not actually be as strong or as in shape. Suddenly I realized that my resistances had diminished drastically. I didn't need quite as much "shape" because I cooperated with rather than fought my physiology. At the minimum, I observed if I could not cooperate. No thoughts or feelings negative to following Hal occurred, no introspection, no conflicts. I let my genes take control. My forefathers had always been frontiersmen and therefore special forces, multi-specialists.

A powerful feeling swept through and lightened my entire physical body. I had returned home. Hal was my Uncle Polk who had submarined into Tokyo Bay in 1942, my cousin Bobby Lee, first rider of Brahma bulls in the county rodeo, my grandfather who'd survived one of the deadliest Texas feuds, my great-grandfather who'd unilaterally seceded from the seceded confederacy, and had dared the Grays to take him at his Nacogdoches Ranch. My father who had given me the family long-barrelled flintlock at the age of twelve, made in eighteen twenty-two, along with tales of the frontier horses whose smell could distinguish friend from foe. My Uncle Jack who had commanded the last American cavalry unit and I had seen the horses stamping in the stalls on the bitter January morning at Fort Riley in nineteen forty-two. The mounted cavalry lined up for morning inspection just before being converted to paratroopers, Jack telling us later, "I lost eight hundred out of the thousand in an hour in the trees where we landed right by their tanks hidden in the forest. You have no idea what it's like, watching boys you've trained dying all around you." And the top of the paratroopers becoming the special forces.

The giant parabdha karma wheels of generations of accepted lore turned in my head as Hal stubbornly, silently, at times inspiredly, covered clicks of rugged country around the special forces base. It became past competition, onto where we showed each other all our special tricks, all without a word, how to never lose a contour line when going up or down, how to accelerate going up, and how to slalom run however so slightly going down, how to alternate looking at the ground beneath, the opening ahead, the great tree that would landmark the next hundred feet, the line of rock outcropping the occasional glimpse of the ridge line, how to spring out of the mud before sinking down by means of a fast foot vibrato, how to catch an unspiked tree trunk to accelerate a change of direction on steep slope. My movements began to develop into the comprehensive pattern flow I'd recognized socially with Dinka, emotionally with Anil, and intellectually with the Tibetans. Hal and I were moving and no six hour dancing spree at a crossroads "country club" could touch that twenty-five click joy-ride. So I had returned to America here in a tropical jungle halfway around the revolving, rotating, wobbling, spiraling, evolving planet, after having left it a year and a half ago on a Yugoslav freighter out of an obscure Brooklyn berth. Five hours later, Hal and I loose-stepped back into base.

"He'll do OK," he told Jackson.

Backing up the Special Forces, a hundred Montagnard helped patrol the area. We left on patrol the next morning, Jackson, Hal, Jones, myself and thirty Montagnard. Four men carried a thirty-caliber machine gun plus ammunition, the rest of us our rifles and some grenades.

The one-man-at-a-time trail led through bamboo groves and a classical three-tier rainforest. Damp, with only the occasional splatter of sunshine. A veteran Montagnard and Jackson led the point. I found myself placed in the middle with Jones. Hal brought up the rear with another veteran Ko Ho.

Mid-morning we stopped for a break. The feathering three inch thick bamboo towered above us, silent and motionless. "Dakang", Jackson called.

An unarmed man with a backpack appeared. Jackson broke out a beer for the three sergeants and myself. "Any fool can be uncomfortable."

The men sipped water from their canteens. Ahead and behind two men crouched, on guard. We looked at the topo map. The old mining engineer training snapped in. I showed Jackson a special trick in map orientation. Suddenly we were a team. His finger traced our route.

That night, after covering nearly thirty clicks we camped on the steep slope of a mountain. "Difficult to ambush," Jackson said.

The men quickly slashed bamboo and lashed small beds together. Other joints of bamboo had one end plugged with clay, rice and water poured inside, plugged on the other end and put over the fire. When one of the plugs exploded out, the rice, cooked au point was dumped onto the field plates and eaten.

After a dreamless sleep and breakfast washed down with cups of tea, I helped cover the camp's traces and we continued an alert pacing down the cool, shadowed trail.

Suddenly we reached a rise on which only tall grass grew. On an arm motion from Jackson we formed into a line and lay on the top of the ridge looking down on a grassy valley through which we could hear a stream running. In front of us, a hundred yards away stood the vertical log walls of a stockade. My heart pounded like a mechanism in my chest. I wasn't tired at all and I could count it by its sound, one hundred and three times in sixty thousand-and-one counts, forty-three above my resting count.

Jackson crawled along the line. "Stay here, ready to back me up with covering fire if I give the signal. I'm going to check it out to see who controls it now."

Jackson stood up and strode slowly down the hill toward the entrance in the stockade. The machine gun stealthily set up, we all kept our rifles at the ready.

My mind flipped back fifteen years. Herb March walking through a mob of white Irish surrounding the Chicago southside home I lay crouched in. The bricks littered the porches, front and back, some lay inside the broken windows. We had invited a black friend over for dinner and now had been surrounded for a day and a half. Herb in his conservative sport coat and blocked hat, a hero of the steelworkers organizing committee of nineteen thirty-seven, walked right up to the door, and walked in as we swung it open. An hour later he walked out right through the mob, looking neither left nor right, never speeding up or slowing down.

Jackson continued his deliberate lonely walk in the silent heavy sunshine. In all the scene he alone moved, not even a hint of wind in the leaves above or the grass below. At the great gated entrance, he rapped three times with the heel of his fist. He waited, and rapped again, three times.

The gate began to open slightly and a rifle glinted. I touched my trigger with the inside of my finger.

The gate closed again for a minute, but Jackson looked relaxed. Then three men in their loin cloths came out and Jackson and they exchanged embraces, Jackson towering nearly a foot above them. He waved at us and we all advanced cautiously down the slope, the two sergeants, each with his Ko Ho buddy, covering both flanks till we entered.

Inside we saw the tunnels, the escape under each house. "VC attacked twice, when they attack all the women and children go below," Jackson said.

"In the States they say only the VC have the tunnels to go into to avoid American attacks."

Jackson spat on the ground and gave me a looking over. "Why do you think we said OK for you to come along? So as you could see something, maybe?"

That night we feasted on pig, sweet potatoes, and dampay beer, self-fermented from the rice. An old woman sitting at the edge of the party made sure we drank to the bottom.

Jackson left the village some ammunition and grenades. "Little enough we can do for them out here. But they're determined to fight."

Late that morning, single-filed on a steep mountain about a hundred feet from the ridge, a sharp crack. Several cracks followed immediately thereafter. Men dived to the ground, in front and behind me. My body hit the ground, independently of my will, a fraction of a second later, stretched out flat. The sound of firing became continuous but it only slowly became loud to my ears. I saw a body had fallen on the downslope side of the trail. We all faced uphill, but everyone else but me had fallen down on the upslope side of the trail.

The trail was a foot and a half wide. It looked as wide as a highway. Now I knew why they fell on the uphill side. The firing continued unabated. I felt I had to go across that trail and in line with the others. Nothing had ever seemed so exposed to me as that opening. Finally I managed to lurch across.

To my surprise I found myself ready to shoot. I would have killed if I had seen anyone coming, but look as I would I could see no one, nothing but tree trunks and blobs of bush and spats of sunlight. The Ko Ho to my left and right had already shot off their third clip. The firing suddenly stopped. I made questioning signs. The Ko Ho indicated none of them had seen anyone.

The point man with Jackson had been head wounded by a sniper. "I emptied a clip at him," said Jackson, "but I missed. He got away over the ridge. He or they won't be back now. We've got to get Manang back. He's wounded badly." Jackson had bound up Manang's head. Two men had got on the stretcher. We started back to the base.

In the middle of the afternoon, we were passing through one of the rare natural clearings. A chopper clattered overhead. We started waving our rifles and caps in the air.

Everyone gathered to attract the chopper's attention so it could fly out the wounded man. The chopper made a long turn in the sky, and came down low at us, several hundred yards away. We cheered. A sound like giant raindrops hitting the ground nearby.

"Fifty caliber! He's shooting!" Jackson waved more furiously. Everyone waved. The raindrops grew to a storm. Some of the bushes cackled.

Suddenly I saw three bodies stuck behind a great tree in the middle of the clearing, and my body sailing to land on top of the three beneath, another landed on top of me. We oriented exactly in a straight line with the approaching chopper. The plop plop plop was all around us punctuated with thuds into the tree. The Ko Ho had scattered to the clearing's edges. The chopper turned over the forest and skimmed back over us, firing more rounds. We had reversed to the other side of the tree. As it flew past we ran to the clearing's edge and Jackson once again tried to wave at the chopper. I put up my rifle to fire, ready now to bring down the Americans trying to kill us. Hal pushed my rifle down. "You can't do that." "Why not, they're trying to kill us." "They're Americans." "Does that mean we have to helplessly die?" But I swallowed the words in my throat. It obviously could mean that we had to die if they improved their aim.

The chopper roared at us again as Jackson joined everyone clutching earth behind whatever tree we could find. A noise of great branches breaking, falling around me. I looked up. Rockets that looked like little toys were falling to the ground together with the branches they had pruned. The machine gun bullets sounded more than ever like rain falling through the thick layers of leaves in the triple canopied forest.

The chopper's sound diminished. "Ran out of ammo," said Jackson, "off to get the A-26's and napalm us. They'll burn the whole mountain burning us."

"Never seen anything like napalm," Hal said shaking.

Napalm, everybody saying napalm. Caves of burned bodies from World War II movies. But we were Americans. So what. We would be napalmed by Americans. Panic struck.

"We've got to get five miles from here in an hour," Jackson shouted. "You," he yelled at me. "You're a big guy. You alternate on the stretcher with me and these two guys."

We started running down the trail. Panic created powers on the level I had experienced in the Nile, in India, in Tibetan Nepal.

Nobody tried to stop themselves or anybody else. Nobody stumbled, nobody got lost. Already panic had kept casualties to zero from the chopper's three attacks. Adrenalin in flight could last longer and flow freer than adrenalin in fight because it didn't meet any opposition. No obstacle existed to its full manifestation. Panic had a bad name because of fires in theaters. In the wilderness, with skilled men, in front of overwhelming force, panic became a life saving power. No one tired. I gave myself luxuriously up to the panic, it seemed everyone did. We ran, thirty-three healthy men, carrying one wounded, as a single unit, a panic, no one faltered.

Over the mountain, down the mountain, over the next hill, down its rugged slopes, a deep stream about fifteen feet wide suddenly appeared threatening our escape.

Jackson plunged across. I followed. I or rather it ran across. On the other side a quick glance showed my pants sopped only to the knees. Had it run so fast that it almost walked on water? My rifle was dry. The panic began to subside when we were two valleys away from the threatened hilltop.

We gathered around Jackson. He broke out the radio and contacted base to fly the wounded soldier out. Then we made it back to base. "We've gotta show our presence every month,"

Jackson said.

Four days later, I started to reread a vignette I had just finished about a scene that occurred on returning to Saigon:

> Except for the war we might just as well have been sitting on bar stools back in his coal mining hometown in Kentucky. The beer was cheaper, more girls were hanging out at the bar and their lips were slightly fuller, their noses shorter, and they wouldn't freckle in the sun.
>
> But there was the same expectant Saturday night air, the beer flowed down (no moonshine here), and the same need to talk about his trade, his wife, his hopes.
>
> He was profane like every fighting soldier except General Lee. He happened to sit down by me, and in war propinquity makes for quick reaction.
>
> "I'll buy your next beer," he said. "You look new."
>
> "Just got back in today."
>
> Then I confessed to a gutless feeling and started asking questions. Where was it safe to go now? Was it true you couldn't go out of Saigon at night? And suchlike.
>
> "Alright to go up the river to Cholon Chinatown long as you don't go too deep into Cholon. Nobody'll bother you here in the City."
>
> I remembered the blasted-out windows on the fifth floor of the Caravelle Hotel down the street but stayed quiet.
>
> "Only seventy-six days to go. My year's up then. A year's a long time when your wife's nineteen."
>
> He, it turned out, was twenty-two. Finishing his third year in the paratroopers. "I know what you mean by gutless. Hell, I been too scared to move. They got me here." His finger moved along his side.
>
> "They sent me to the Philippines. Officer was there I knew. Damn fine man. They put two clean bullet holes in him and they fixed him up good in Saigon the same day. Looked clean as a whistle. But they sent him to the

Philippines. They knowed the infections'd come. Sure enough the twenty-second day he swole up. Broke out everywhere."

"My daddy was a master sergeant. I reckon he'd be proud of me now. I'm a E-Five and I reckon I'll make six this next year. I'm gonna put in for Special Services — that's in fifty- four weeks you know, one year and two weeks. They want you trained. They got spit and polish, but I can make it. I learned how to make them old boots stand up and shine."

"Special Services was President Kennedy's own idea, I mean it was his baby. Yeah, they're good. I want to tell you that Viet Cong is ee-leet, too, the ee-leet of the ee-leet, they're every bit as good as the Special Services, maybe a little better because they been at it longer. These fellows with us, now that's a different matter. Course, I'm just a old paratrooper, but that's how I see it."

"You know, I ain't goin' to buy that rubber tonight. I wanted to talk it out some. You know, you got to have some when you're over here a year, but I want to be back with my wife. I've bought her lots of surprises. She's goin' to really roll her eyes. Hell, what I really wanted to do was talk it out. It's goddam good to talk."

"There was a fellow in my outfit, he got killed coupla weeks ago. He was a real good fellow, he was my friend, but I guess I think of him as kind of a hero, yes, I guess do. I mean when they got him he'd used up all his ammunition."

"Master sergeant's best. Sergeant-major takes it from everybody. I reckon they ain't nobody gonna keep me from bein' master sergeant. I figure E-Six this year, then Seven, then Eight, yeah I'll make it."

"Some of the fellows say I don't look out for them enough, well, I can't wipe their noses for 'em. I'm at it all day tryin' to get our job done but I'll see two of 'em promoted to Fours before I leave. One of 'em's a cruddy bastard, a real slophound, but he's good, I'll put him up for it, he'll have a good chance. Hell, if he don't make it, it'll show him not to be so cruddy."

The nightly rain. Curfew and the white-uniformed police patrolling all the streets. We separated to hit the sack.

The next day after another encounter I wrote another one.

The man piled up at the bar door in planes and cubes, a Grenade, Hand, Fragmentation, ready to explode.

He joggled, safety on, to street level, strode past the squatting wrinkle-faced seller of smooth sliced papaya, and rigidly turned left. Toward me his jaw jutted steel blue prickles.

"This street's not big enough for both of us."

"I said this street's not big enough for you and me."

"Because I say it's not."

"Because I don't like your looks."

"Mister, I'm looking for trouble and when I saw you I knew I found it."

"When I beat a man up, he's beaten like he's never been beaten."

"Because it pleases me."

"Alright. I want some action. They won't send me to the action. So I booze here looking for action."

"Yeah. I drank twenty-eight beers one sitting. Something, huh?"

Below checkered Bermudas, his square knees and slablike calves foundation-stoned him where he stood. His clenched fist measured my jaw. He was half-drunk. That small non-coordination would be my only chance. Except retreat.

So why did I eyeball to eyeball this ticking bomb? Like was it cool not to leave this spot of 15 piastre beer and 200 piastre girls?

He held me. My left knee shook, a leaf, but I'd just hiked hours lonely, restless, along glistening tidal flats, banked freighters smoldering under heavy stars — once more as in a thousand and thirteen before places seeking a new concatenation of the secret. At last at least some action. I could not retreat.

"If you don't start moving in ten seconds I'm going to break your jaw."

Meaning exactly ten seconds or a minute? My exact response never entered memory, but softly to the effect he could take all the street except the inches where I reassembled and the little air I carbonized. Perhaps he didn't want to hit a man who kept his hands down. Perhaps he had only wanted to make me interested. Perhaps — but one may read many books, travel around longitudes and latitudes, and still one's own behavior, not to lie of others', though examined, remains enigma.

"You only have two things I want — your beard and your words. The army won't let me grow the one and I quit the tenth grade."

He plucked out, aplomb, his pack of cigarettes and offered.

"I only drink."

"Okay, I'll buy you a drink."

He plunged off across the street to a bar mainly for officers. All the girls on that side wore the graceful Ao Zai and long hair, on the other side slacks, short skirts and bouffant hair. But officers and men garnered as often the same bad luck.

He turned on his stool to face me. I flipped after a second's confrontation. Try looking at another man. Try how long. If you're like me, pretty soon you'll tap your fingers, cross your legs, look at his shoes, if no one else around you'll look at stains, a slab of sky, start swilling. Like when you tune in someone's wavelength there's buildup, the waves faster, the sounds shriller, finally it's pain and you turn the damn thing off. Except with ones like Anil and Chini Lama.

Turned on his stool, he poked his bristler ever forward. It was like the side of a house falling down and inside a family screaming.

"I'd give this much of my dick to express myself." He placed his thumb and index finger two inches apart. Maybe you got eight and can afford it, my associations automated.

"A man can be good at anything, but he can be great in only one thing." He repeated himself.

"He can be great in only one thing."

I sat there like a Greek before the delphic. Where did his sentences come from? Magazines, street corner bull, hand-me downs from atheist uncle, some origination of his own flashing electric clickers?

He'd finally landed his punch. Jack-of-all-master-of-no-trades me. Why hadn't I become a great pooh-bah or blah-blah?

"A man has to find that one thing he can be great at."

How many times I'd tilted this tournament. I know all shocks of non-recognition. Yet I listened like baby listens to rolling who knows-what-they-mean mama papa speeches feeling must understand because it might be magic. What? What is one thing I can be great at?

"The one thing I can do is fire weapons. At a hundred and twenty-five yards I can empty an M-14 into a space the size of my palm. There's only one thing each man can be great at."

"Firing weapons, outthinking the enemy that's what I'm great at. But since I haven't really done it, since they stashed me in maintenance, how do I really know?"

How does one know? He's beating me up worse than with his fists, only deep I love this way beaten-up, physically no, no, a broken nose too permanent, only one body, but have a few thousand souls to spare, expendable, flay them again and again into the fray, meat hook each new faith on questions. If you don't do it, how do you know? But there're a lot of things to do, and what if you pick wrong five times running, life's gone?

"We used to throw cards. Face down, we fight, face up, we didn't. Stupid, huh? But that's the way we did it when I was thirteen."

A tough one, alright. I saw him by the pool table on Halstead Street, cockier than the rattling street car, colder than the Lake Michigan wind, more hood than the stockyards stink, bending, stroking his cue, carefully rolled to check the deviation, lining up on veed thumb and finger, then follow-through breaking the rack wide-open with a blasting bust, transmuting his grin into swaggered one hand lean on table while at least three balls plunk plunk plunk. I saw the time he found out you don't feel pain when you hit the sidewalk during a fight, only later, and that if you've won, that stiff bruise's nothing to the burned-out peace surrounded and supported by respect and self-respect. And the time he found out he had to prove it again — and again.

"This Caribou bellied in with a hundred and fifty-five gigs, you know what I mean? I'm a leader. I took that ship over. I owned that ship for four and a half days and then they flew it away. You never own anything working with aircraft. You fix 'em up and they fly 'em away. You don't ever own 'em."

"Have you ever seen a sergeant who was a leader? I'm an idolizer. All Americans are. I idolize my SP-Six. He works all day on a prop and doesn't catch a speck of dirt. I worked a half-day once, from seven to twelve, and I didn't catch a speck, working with grease. My front buckle shiny as when I started and the back of my belt clean. I'm serious. A soldier should be clean."

"Yes," I said watching myself grow as formal as a placard.

He had begun to maunder.

"There's that one word we don't know. You tell me the one word to help suffering. You tell me the one word to tell someone whose buddy's been killed. Who's been shafted by a Dear John letter. There isn't any word. I'd give anything to know that word. Only poets are happy. They come closer to knowing that word than anyone else. You're a poet, aren't you?"

I wanted to say yes, yes. I wanted to face the delphic again. The Dear John bit pissed me off. I couldn't picture it so suffering. Who wants a girl who doesn't love you? I didn't want to empathize. I wanted him to stir old fear, arouse awe by hurling brute words charged with lightning. He was lying. Suffering would be his being wounded, not a buddy's being killed, not his being killed because that's the end to suffering. Why didn't he admit he was afraid?

"I think I'd kill any man I saw smirking at the American flag, even another American."

"I think you might."

"You know, I could hate you all over again. I'm supposed to be a big tough motherfucker who never gives. Why am I talking?"

He stared at me but I had disappeared from his attention now focussed on unaccustomed introspection. He swung around, slid off the stool, and bounced away. We'd've had to get drunk, or laid, or in a fight or all three if he'd stayed on.

After those two scenes, and after studying closely the two vignettes in which I had written out the scenes, I knew that not only had I escaped America in Tangier, but that by now I had found at least enough of the truth to allow facing my culture conditioning objectively while remembering my own being. I must now go back to America and take up regrowing my being where it had been truncated, pruned into a bush instead of opening to the stars.

This new aim only gradually became clear to me and then

compelling in the months that followed the panic on patrol in Vietnam.

I left Vietnam in April, nineteen sixty-five, after Johnson had ordered the bombing of Hanoi and the introduction of regular infantry. Clearly the Kennedy ideas of a noble elite fighting with intelligence and ground truth, mingled with the population, competitive fishes in the ocean, had disappeared and a Texan figurehead would order into action the quantitative attrition warfare of World War Two setpieces designed to destroy fascist legions ordered to advance killing until killed. Against a guerrilla-war trained ideological army whose objective was to gain control of a limited amount of territory, and trained to retreat and disperse at the first hint of superior force, especially one given sanctuary along a six hundred mile frontier line, this type of warfare was stupidity squared. Nothing more of any interest could occur in Vietnam, only casualties, until one side bled dry. Infantry officers, paunchy and paper loving, arrived to take command of the special forces.

"Why's this guy out here?" shouted Captain Louis Miller, newly arrived to take command of the eight sergeants, just the day after our patrol ended. West Pointers now wanted to get in on the glory of the Special Forces. He had flipped, seeing me at the base. "Get him out of here, now, Jackson. These writers should stay in Saigon. We don't want any of 'em out here, you understand? We can't be responsible for them. What if one of 'em got killed? What would they say then?" Miller strode off. Jackson looked at me. We shrugged.

Flying out on a wangled seat on an airforce plane from Da Nang, bustling with the booming new port being built, "to gain strategic domination of the South China seas", the Americans forbidden to leave base and fraternize with the locals, on the way to Hong Kong. The commander of a Mekong River gunboat sat next to me.

"You poets," he said, "you artists, and don't tell me you're not, feel separated from us, but we feel, too. I read Dostoyevski in my off-hours. Now I'm flying a blond for a thousand dollars to Hong Kong for a five day R and R. I know we have to put it all together, the establishment, art, romance. So why don't you and I start, right now?"

I admired his bravery, taking that boat up and down the ambush-laden Mekong, peering into the bush, the enemy necessarily having the first shot, just like on my patrol, except he did it again and again. At the same time I knew that note in the American voice, defensive, self- justifying, blocking any interchange that might introduce the completely new, setting up the conditions of emotional stasis while pretending openness intellectually.

Dostoyevski. It was symbol talk. It meant I couldn't mention Jarry, Artaud, Burroughs, my own writings, above all it meant crime and punishment, not notes from the underground or the idiot.

"It's a shame we don't have more time to talk, to see how much we have in common."

"That's right," I said. A few weeks later I heard he'd been ambushed and killed a month after his return to the Mekong.

I retreated to living on board a junk among the boat people just off Kowloon. I helped do medical work there to pay my room and board, watched the wedding junk and the ice junk plow their rounds in the watery community. The masses of refugees from Maoism poured into Hong Kong. The hillsides blossomed shacks. The junk's manager advertised for a doctor and fifty-two qualified Chinese M.D.'s who had left the mainland in the past two months applied.

In Japan I met no one except a dead poet.

Sitting on the wooden walk gazing at the raked sand waves ceaselessly in the motion of motion whitely besieging the dark peaks that rose infrequently from the ocean, Basho's haiku ceaselessly annihilated all my arising thoughts,

The pond
A frog jumps in
Splash!

At the arising of each thought for three days this version of his poem struck, sometimes viciously, sometimes laughingly, sometimes indifferently, to kill that thought before it hardly got started.

The pond
A frog jumps in
Splash!

The pond became large, still, deep, one frog after another jumped in, each splash evaporated into stillness, the frogs crept out at the pond's margin, and prepared another short-lived splash.

Would all the experiences of Tangier, the Nile, the Subcontinent, Vietnam, all transcendent with possibility vanish like a series of frog splashes? At last I saw the frogs and stopped and started them before they splashed, and finally I saw the pond and kept the frogs out, and then let them go, jerkily till they splashed, and let them climb out again and rest before the next splash. Finally I became a frog breeder and trainer. Splash! Splash!

I remembered talking to an eight year old ship's chandler in Tetuan, Lataif the subtle. Lataif held me enthralled for an evening telling me the adventures of his world-traveling life, the great ideas, battles, fortunes, and teachers he had encountered. We kept refilling our hot mint teas. At the end he looked into my eyes for several minutes. "And what remains of all this, my friend? Three hours of words about events that I find ever more difficult to remember. It has all passed by like a dream. My life has been a dream, unreal. My life, which fascinated me for years, and whose story can still fascinate you for hours, I must judge to be a waste. And I have no idea how I could have done anything better."

I had never seen a sadder man, a man who had done everything that can seduce with glamour, but who had lost it all to time, the little percent the house takes on each throw until the gambler loses all.

In Kyoto I realized that I could not settle for being Zen, a snowflake emerging into crystalline splendor, and then melting, happy to be even though functioning as a vanishing snowflake. I refused to accept the apparent as unreal, doubtless the apparent was only a part of reality but at minimum, I saw, the apparent must be not only the bridge that led to reality, but part of the arrival turf itself. And the most apparent structure in all the world to me was history. The universal history, the biospheric history, the cultural history, my history. Everywhere I saw choice pregnant with decision, ready to deliver the future. I needed to conquer time, liberate space, and upgrade energy to achieve reality, to escape the world-abdicating wisdom of Lataif and Zen, the ship chandler and the snowflake.

Suddenly I decided to return to where I had begun to journey around the planet, Manhattan. A certainty told me, that's where you must fight your next battle to find your way.

The Real Cafe

BACK IN MANHATTAN in the late fall of nineteen sixty-five, I spent hours each day in a new café, that is, it wasn't there when I'd left in October nineteen sixty-three, escaping to Tangier.

Prince's was run by tall tough shaven-headed Prince, an Italian, who had a tall lean flashing eyed spectacular black girlfriend. Wooden floor, glass windows, perfect lighting that could support reading, but not one bit more, solid tables, comfortable chairs, space between the tables, great coffees, steaks, and desserts. A half block north of Bleecker Street on Sullivan, it centered the hip vibrations of the South Village as some had started to call the area below Washington Square. Bill Layton and I had set up in a loft south of Houston Street for our chemical experiments about three and a half blocks away. I spent all my time between Prince's and the fourth story top loft, and a small rent control pad on West Broadway that Bill passed on to me when he moved in with his redhaired green eyed bar hostess in the new apartment house at the end of Bleeker Street. Below the loft was a three story warehouse run by the two lean and wary Feinstein brothers. We paid a hundred and fifty a month and it was illegal to sleep there, but I often did. The Feinsteins were strictly business, and never bothered us except for a once a week inspection.

I had made up my mind that what I needed after discovering why and what was to contact a group who had access to how, to ways of developing being based on more comprehensive and deeper knowledge than available from individuals, however advanced, and I had decided on pursuing three routes to achieve this aim of learning how to learn now that I had a definite taste of how much there was to learn.

For the first of my three routes, I spent eight hours a day at Prince's, watching everyone who came in, listening avidly. Since I had decided that this place would be where men with knowledge could meet, I had to be there to recognize them when they came in, as they probably would, perhaps only once. In addition, I had to stay awake because men of knowledge probably had better camouflage than a Special Forces sergeant in the jungle. At Prince's I also followed my second route, making my way by vibration choice through the diverse volumes acquired weekly at Weiser's Bookstore, a semi-scholarly method I used in order to be ready to identify a genuine miracle, by learning to detect error.

Eight hours a day, for my third route, Bill Layton and I worked at our laboratory on ibogaine, mescaline, and dimethyltryptamine. Not for money, but for our researches into upgrading physiology. We did not use any psychological techniques or look for any results at all in that direction. All of those people left a curious taste in my mouth, as if they were looking for great buys in a used car lot. For cash we sold special fluxes to join the tiny wires that did the work for space and war and finance by transmitting rapid information. They required a truly dangerous chemical, hydrazine, unlike the shaman-tested-for-generations substances.

Unwilling to rely solely on my observational training and ability exercised at Prince's and upon finding a complete explication in the debris of systems and mystics whose pulsing fragments filled the books I devoured. To make surer the finding of my way to the needed knowledge, I endeavored to expand and intensify each of the four levels of consciousness that I could identify by means of various exercises connecting them directly with the substances we produced. Dimethyltryptamine taught me the ultimate tempo and epiphany of colors, no second trip ever needed, amazement at phenomena could hold no attractions beyond these demonstrations, but as far as knowledge went, its application would stop at art; ibogaine introduced me to the supernal stillness in which ongoing nature treats man like harmless stone and drops its disguises to reveal its actual processes to entranced sight and hearings and I saw how the forest yogis with prana and datura must have mastered their cosmic perspective but that as far as knowledge went, its application would stop at philosophy; mescaline that accelerated my splashing in the pond to Mogul fountains of thought, and I saw the power of the formless, and from whence all the forms could be generated, but that as far as knowledge went, its application would stop at the ceaseless world of the love of beauty of the mystic's union with the creative force. The fourth level, moment-by-moment contact with a continuous integrator of the world of forms, of the world of manifestation, and of the world of the inexhaustibility of phenomena, came dismayingly seldom and not for very long. I did a two hour workout each day counting breaths and movements, sweating, to survive with health these weekly sessions plus the reading, the observation watches at Prince's, and the detailed work at the laboratory. I learned to work my way out of many a tight corner on those vigilant journeys when it seemed some organ or muscle must have reached a point of no return.

One day at Prince's, early in January 1966, three men walked lightly and decisively in and dropped gracefully into the seats at the table next to mine. They ordered T-bone steaks and coffee and their three faces were the most interesting three faces I had ever seen together. One black, broad, handsome, vital, passionate; one thin with coal-black eyes laughing phenomena to attentioned order like a whip cracking in a circus ring; one brooding, melancholic, seductive, with the aching wit of inner bite.

They were talking knowledge, knowledge about which I would give whatever was demanded to gain experimental experiential evidence. I could not make out the words but I knew they were talking the hermetic knowledge, "locked with the seven locks." My journey, my reading, my experiments made the scene as clear as moonlight. A scene in which I was as yet only in the audience, but at least it was a scene from the drama I wished to see, the drama of reality, and at least I had became an audience, and the audience itself received power if not knowledge from the performance. No three men so intelligent and interesting could talk to each for so long, more than two hours, about anything else other than reality with such unbroken attention, lack of argument, absence of sexual innuendo to which I was excruciatingly sensitive, alive backs, clear eyes, wonderful appetite, and evident mutual regard. In short, I fell in love and realized that it was not in love with a projection of myself, not with an individual, nor with a style, but with a group who had been transfigured by knowledge and who were continuing the transfiguring process. They were not carried away by it, not out-of-themselves, seduced. They were operating themselves and the group as I had operated a mine and a chemistry lab. They knew what they were doing to themselves, and why they were doing it, and they were making what they intended to make, and there was no psychology in it, rather a metapsychology, only a direct conscious application of knowledge to the total physiological system, individual and group. Now I knew why Prince's was different. These men, and perhaps others who knew them, met here on occasion and Prince's soaked up their spin-off just as our chemical laboratory had acquired colors, shapes, and mystery that affected our most obtuse visitor.

Two of the men left. The black man came over and sat down in front of me without any nervous facial flutters, arms bent into harmless angles, or shuffling feet.

"You wanted to hear what we were saying."

"I wanted to meet you."

"Well, now you have."

"I have traveled around the world, met great teachers, studied, and explored physiology with breath and special substances. I know the mind is filled with splashing frogs and even how to catch the frogs and play with them. I know how, by confrontation, to measure the relative being of men, and I know that only the remembering of death makes me desperate enough to continue my search."

He looked at me gravely, neither inviting me to continue nor to stop.

"I have dreamed and even made plans to start a group, called Catalysis, which would form a planetary study and search for the truth, within our bodies, within societies, and within nature. I call it Catalysis because we would have to remain unchanged in the midst of the prodigious changes that we would unleash in ourselves and in our bodies. And these changes, not to become meaningless, that is to say self-cancelling phenomena, would have to be aligned with history, so our first step would be to ascertain the objective truths of history."

"Little boy," he said. The zodiacal procession of interesting glitter stopped for me then and there. This man was no older than me, perhaps even two-three years younger by census count. Yet I knew he had handed me this term, little boy, as a piece of knowledge, completely non-judgmental. The first time in my life I heard the truth about myself. Little boy. Indeed for me that was the truth.

I behaved as a little boy, my metabolism life I heard the truth about myself. Little boy. Indeed for me that was the truth. I behaved as a little boy, my metabolism operated as a little boy's, my enthusiasms, fears, ideas, were those of a little boy. Suddenly I saw myself, a little boy running wild around the sunflower edges and scrub oak and quicksand margins always on the lookout for an Indian to play with. Tears coursed down my face as if by their own accord.

"Little boy," he said gravely, kindly, definitively, "that group of which you speak of starting exists already. It has existed since the beginning of the history of man. Be here tomorrow and we will talk some more."

My journey around the planet had ended, three blocks and two and a half years and many encounters from where I had decided to start. And twenty-three years younger. A little boy of ten waved at Prince as he glided in. Prince always alert and cautious from doing time gave a small nod. His gorgeous but cool black lady finally smiled broadly at me and dropped her mask but it bounced back on her face before Prince, smelling something, had whirled to check her out.

The little boy jumped up and skipped down the street and on around Washington Square Park. He was going on a real trip at last. He waved at the Bronx Bagel Babies, he waved at the Manhattan A-trainers. He saluted George Washington. He bought himself a chocolate ice cream. His life lay before him and where it ended, he hadn't the faintest pond for the faintest frog to make the silentest splash. Little boy knew that he was unknown and unknowable by father, mother, mind and world. I stood, my body licking my ice cream, like a cat, I knowing at last that I was invisible to the world, but watching myself in the world, and watching the world around myself, I became visible to myself as myself became visible to me, and me, myself, and I, we strolled around the fountain in the middle of Washington Square, laughing because the world was also and forever young.